ENGINEERING AND ENVIRONMENTAL CHALLENGES

TECHNICAL SYMPOSIUM ON EARTH SYSTEMS ENGINEERING

National Academy of Engineering Annual Meeting

October 24, 2000

NATIONAL ACADEMY PRESS
Washington, D.C.

NATIONAL ACADEMY PRESS • 2101 Constitution Avenue, N.W. • Washington, D.C. 20418

The National Academy of Engineering was established in 1964, under the charter of the National Academy of Sciences, as a parallel organization of outstanding engineers. It is autonomous in its administration and in the election of its members, sharing with the National Academy of Sciences the responsibility for advising the federal government. The National Academy of Engineering also sponsors engineering programs aimed at meeting national needs, encourages education and research, and recognizes the superior achievements of engineers. Dr. Wm. A. Wulf is president of the National Academy of Engineering.

The interpretations and conclusions expressed in the papers are those of the authors and are not presented as the views of the council, officers, or staff of the National Academy of Engineering.

International Standard Book Number 0-309-08415-6

Copies of this report are available from National Academy Press, 2101 Constitution Avenue, N.W., Lockbox 285, Washington, D.C. 20055; (800) 624-6242 or (202) 334-3313 (in the Washington metropolitan area); Internet, http://www.nap.edu

Printed in the United States of America
Copyright 2002 by the National Academies. All rights reserved.

THE NATIONAL ACADEMIES

National Academy of Sciences
National Academy of Engineering
Institute of Medicine
National Research Council

The **National Academy of Sciences** is a private, nonprofit, self-perpetuating society of distinguished scholars engaged in scientific and engineering research, dedicated to the furtherance of science and technology and to their use for the general welfare. Upon the authority of the charter granted to it by the Congress in 1863, the Academy has a mandate that requires it to advise the federal government on scientific and technical matters. Dr. Bruce M. Alberts is president of the National Academy of Sciences.

The **National Academy of Engineering** was established in 1964, under the charter of the National Academy of Sciences, as a parallel organization of outstanding engineers. It is autonomous in its administration and in the selection of its members, sharing with the National Academy of Sciences the responsibility for advising the federal government. The National Academy of Engineering also sponsors engineering programs aimed at meeting national needs, encourages education and research, and recognizes the superior achievements of engineers. Dr. Wm. A. Wulf is president of the National Academy of Engineering.

The **Institute of Medicine** was established in 1970 by the National Academy of Sciences to secure the services of eminent members of appropriate professions in the examination of policy matters pertaining to the health of the public. The Institute acts under the responsibility given to the National Academy of Sciences by its congressional charter to be an adviser to the federal government and, upon its own initiative, to identify issues of medical care, research, and education. Dr. Kenneth I. Shine is president of the Institute of Medicine.

The **National Research Council** was organized by the National Academy of Sciences in 1916 to associate the broad community of science and technology with the Academy's purposes of furthering knowledge and advising the federal government. Functioning in accordance with general policies determined by the Academy, the Council has become the principal operating agency of both the National Academy of Sciences and the National Academy of Engineering in providing services to the government, the public, and the scientific and engineering communities. The Council is administered jointly by both Academies and the Institute of Medicine. Dr. Bruce M. Alberts and Dr. Wm. A. Wulf are chairman and vice chairman, respectively, of the National Research Council.

Preface

Every year the National Academy of Engineering (NAE) hosts a public symposium as a part of its annual meeting to encourage discussion among NAE members and the public on topics crucial to the nation's technological welfare. The topic for the symposium at the 2000 Annual Meeting was Earth systems engineering (ESE)—the focus of the NAE's activities on technology and the environment.

ESE is an emerging multidisciplinary area based on a holistic view of the interactions between natural and human systems. ESE addresses global, complex, multiscale, multicycle phenomena, such as climate change, as well as problems of global importance, such as urban design. The goal of ESE is to improve our understanding of these complex systems and develop tools to support technically sound, ethical decisions. ESE attempts to frame problems in a way that maintains important connections yet enables the development of effective solutions.

The twentieth century was marked by unparalleled advances in technology and the development of world resources, as well as unparalleled growth in the human population. Human activities have left their mark everywhere on Earth, and in many places humanity has significantly reshaped natural systems. Our ability to effect changes through technology has grown faster than our understanding of the technical, social, and ethical implications of those changes. Thus "success" has sometimes come at a cost—often unexpected—to the environment, biodiversity, and human society.

The human population has grown to the point that the Earth can no longer absorb and compensate for changes caused by human activities. As our impact on the planet increases, engineers and policy makers must become more aware of the multiple time scales and diversity of environments and populations affected by their actions. As the human population approaches nine billion, we must carefully weigh the costs and benefits (including the social and ethical costs and benefits) of our actions.

With our technical expertise, we can now exert considerable control over natural cycles and systems. However, this control can also perturb these cycles and systems, leading to unintended consequences. These costs can often be reduced if we take into consideration the broader community when we weigh the options for meeting local needs. We must also consider possible "emergent" properties of human and natural systems (properties that were not anticipated in the specifications of the systems) when planning and implementing projects and policies.

With technological advances, especially in sensing, systems management, communication, and information processing, we can now more accurately predict and manage the impact of our activities. As these technologies continue to improve and our experience in applying them to natural systems increases, we will be able to act more proactively in managing the interfaces between the human and natural worlds.

Our challenge for the twenty-first century is to improve the global human condition without mortgaging the heritage of future generations. ESE offers us an opportunity to develop the tools we will need to meet this challenge. This volume is intended to illuminate and frame this debate.

Wm. A. Wulf
President
National Academy of Engineering

Introduction

In the twentieth century, we witnessed the transition of a world dominated by nature to a world controlled, designed, and powerfully influenced by humanity. Through technology, we went from making decades-long changes on a local scale to making centuries-long, global-scale changes to the Earth's biosphere. Advances in technology in the last century supported an unprecedented increase in human population, longer lifespans, a higher standard of living in developed nations, and a broader scope of human influence that now touches the most remote areas of the planet and even into outer space.

Consider some of the benefits of this trend: more reliable food supplies, better disease control, longer life spans, more material comforts, and faster communications and transportation. But also consider the following costs: loss of biodiversity, salinization of farmland, environmental contamination, overcrowding of urban centers, and a growing vulnerability to energy and equipment failures. With our unprecedented technical capability, humans are reshaping the Earth to fulfill our needs and desires, but we are just beginning to understand that these changes have much greater and broader impacts than we envisioned.

When World War II ended, industrial countries were able to capitalize on explosive advances in science and technology to create a plethora of goods and services, not only for the military but also for the civilian economy. We were fast becoming as rich as Croesus. We soon found, however, that we were also sitting atop a rising garbage pile created by our production processes and life styles. *The Affluent Society*, by John Kenneth Galbraith, and *Silent Spring*, by Rachel Carson, were the most influential of many voices that rose shortly after midcentury warning us that the *means* by which we were achieving our *ends* of material wealth were seriously threatening the environment and human health.

Since the 1970s, Congress has enacted laws to regulate pollution and protect the environment, and the science and technology community has devised cleaner ways to provide material "goods" with many fewer environmental side effects. Scientists and engineers have undertaken major research to improve our understanding of the dynamics of the biosphere. In the past 30 years, we have learned a great deal about the functions and frailties of natural ecosystems, including freshwater cycles, weather and climate, stratospheric ozone, ocean currents, and biodiversity. However, we have often remained narrowly focused on solving a problem without regard to the magnitude of the interactions among human and natural systems. As a result, many of our technological systems do not support the iterative decision-making processes necessary to address complex problems. Now that we are more aware of the magnitude and complexity of the environmental challenges facing us, we must not underestimate the difficulties that lie ahead. The often unintended consequences of our technologies reflect our incomplete understanding of existing data and the inherent complexities of natural and human systems.

Earth systems engineering (ESE) is a holistic approach to overcoming these shortcomings. The goals of ESE are to understand the complex interactions among natural and human systems, to predict and monitor more accurately the impacts of engineered systems, and to optimize those systems to provide maximum benefits for people and for the planet. Many of the science, engineering, and ethical tools we will need to meet this enormous challenge have yet to be developed.

Innovative engineering is a key tool in addressing emerging global threats that are caused or exacerbated by human activities. Climate change, loss of species, destruction of water resources, depletion of fossil fuels, and accommodating at least three billion additional people in this century are among the challenges engineers must help meet. The challenges are not only complex in themselves, they are also interrelated, and they must be addressed through global cooperation. Thomas Jefferson's call for institutional flexibility seems more prescient than ever: "As new discoveries are made, new truths discovered, and manners and opinions changed with the pace of circumstances, institutions must advance also to keep pace with the times."

The engineering community faces a three-fold challenge. First, working in partnership with scientists and other representatives of intellectual domains, we must try to analyze and get a clearer understanding of the nature and dynamics of global environmental systems. Second, we must create processes, products, and infrastructures that will enhance our quality of life, stabilize population growth, and assure a healthy, diversified environment. Third, we must work closely with political leaders to develop thoughtful public policies that protect the global commons and enable sustainable services.

The longer we delay facing these challenges, the more difficult our task will become. The NAE symposium on Earth systems engineering, held as part of the NAE's 2000 Annual Meeting, was intended to be a wake-up call to the engineering community.

John H. Gibbons
Chair

Contents

Keynote Address: It's the World, Stupid!   1
Norman P. Neureiter

PANEL I: UNDERSTANDING, ADAPTING, AND MITIGATING CLIMATE CHANGE THROUGH ENGINEERING

Climate Systems Engineering   9
Robert M. White

Improving Climate Assessment   17
Robert W. Corell

How Camest Thou in This Pickle?   23
Jerry M. Melillo

PANEL II: USING BIOTECHNOLOGY FOR THE BENEFIT OF HUMANITY

Genetically Modified Organisms: An Ancient Practice on the Cusp   35
Maxine F. Singer

Benefits of Biotechnology 45
Edward A. Hiler

Earth Systems Engineering and Management: The Biotechnology Discourse 51
Braden R. Allenby

PANEL III: ENGINEERS AND POLICY MAKERS: PARTNERS IN THE DEVELOPMENT AND IMPLEMENTATION OF SOLUTIONS

Gaining a Seat at the Policy Table 59
Anita K. Jones

Successful Public-Private Research Partnerships 65
Kathleen C. Taylor

Defining What We Need to Know 67
Daniel R. Sarewitz

PANEL IV: DESIGNING THE URBAN CENTERS OF TOMORROW

Rethinking Urbanization 75
George Bugliarello

Urban Design: The Grand Challenge 91
Lawrence T. Papay

Hybrid Cities: A Basis for Hope 95
Geeta Pradhan and Rajesh K. Pradhan

Appendix A: About the Authors 107

Appendix B: Symposium Agenda 115

It's the World, Stupid!

NORMAN P. NEUREITER

I was a chemistry major in college, and I must admit that we looked down on engineers. They were a slightly strange lot of guys (no women in those days) with pocket protectors and dangling slide rules. Their building was way across campus, and they had no room in their schedules for English, history, foreign languages, or philosophy. And engineers always seemed to have a kind of academic grease under their fingernails. Admittedly, we chemists had acid holes in our clothes, but we wore them as a badge of honor—because we were doing "pure science" and were pursuing "fundamental knowledge."

My change of heart began when I joined an oil company and found out that chemists were the outsiders whose ideas were usually dismissed by managers who had grown up in the oil patch. Engineers ran the place, turning sulfurous black crude into gasoline and petrochemical feedstocks that propelled the national economy. Later I joined Texas Instruments (TI), where the entire culture was one of engineers. Even the chief financial officer had been a double-E in college. And two weeks ago when Jack St. Clair Kilby, a modest former TI engineer without a Ph.D. and a revered friend, was honored with the Nobel Prize for his 1958 invention of the integrated circuit, my false chemist's pride was dashed forever. So I stand here today humble and respectful in this hall of engineering accomplishment.

Let me start by apologizing for using a rather crude title for my remarks, "It's the World, Stupid!" Actually, I stole this line from a recent op-ed piece in the *New York Times* by William Safire. I did so out of frustration and disappointment that, even as the election campaign reached a crescendo, the vital issues of U.S. foreign policy hardly came up for discussion. One could conclude that the

American people are not interested in what goes on in the rest of the world. Nevertheless, America today is the only superpower and the world's richest nation. The mantle of world leadership is on our shoulders—if only by default. How we exercise our world leadership—and what we do with our wealth and military strength—and how we conduct our foreign policy will be prime determinants of whether this shrinking globe will become a sustainable society, a goal often talked about but completely elusive.

A U.S. presidential election is not just about America; it is about the world. The problems we face—climate change, disaster mitigation, the spread of infectious diseases, safe drinking water, food security, the dramatic loss of species, protection of critical infrastructure, terrorism, proliferation of weapons of mass destruction—do not stop at anybody's border. When dioxin from an incinerator in the lower 48 finds its way through seal meat into the bodies of Inuit people in the Arctic, one sees how small this complex world really is. A Russian cosmonaut said he realized that "we are all sailing in the same boat" when he saw an orange cloud that had formed from a dust storm over the Sahara reach the Philippines and settle there with the rain. The Ebola virus has an incubation period of four days—long enough for a 747 to take an infected person a very long way and to many different countries.

I commend the National Academy of Engineering for its concern about these issues. I strongly believe that our scientific and engineering resources can provide bases for addressing the world's major problems. But I was also pleased to see in the summary notes of the NAE summer workshop a recognition that science and engineering cannot provide solutions to all of the equations that bear on the huge, nonlinear systems underlying our global problems. Cultural, social, political, even religious factors—all of these with coefficients that vary radically from nation to nation—must also be included in the calculations. They could even turn out to be more important than technology.

With the end of the bipolar Soviet-U.S. standoff, we not only have no New World Order. We have a new world of inordinate disorder. Just how disorderly? There are 6 billion people, and the population is increasing by 80 million a year. At that rate, we will have about 9 billion people by 2050. People live in 191 countries, including Taiwan. They speak 3,000 to 4,000 different languages. We can't print world maps fast enough to keep up with the changes. In the business world, megamergers are announced almost weekly as even the biggest companies combine with each other to serve global markets. But in the political world, centrifugal forces prevail. Ethnic tensions and nationalist ambitions continue to divide people, and official sources count some 34 wars in progress at the present time. More than 14 million refugees subsist without permanent homes. Infectious diseases kill 40,000 to 50,000 people every day. The world consumes 78 million barrels of oil every day, and every year we release 6 billion tons of carbon into the atmosphere. Many people are convinced the world climate is already showing the effects.

This is my thirty-fifth day on the job as the first-ever science and technology advisor to the secretary of state. I was happily retired and doing my own thing when this full-time job at the State Department was offered to me, and I just couldn't resist. But working in government seems more difficult today than it was 30 years ago; today there are many more rules, and they are much more complicated. The bureaucracy is more complex, and although we face many of the same issues we faced in years past, they often seem more acute.

My job was created in response to a recommendation in a National Research Council (NRC) study on science and technology in the State Department. The study noted that of the 16 stated strategic goals of U.S. foreign policy, 13 "encompass science, technology, or health considerations" (NRC, 1999). The report also cited specific instances when an understanding of underlying science and technology issues has or is expected to achieve the nation's foreign policy goals.

The primary mission of my new office is to ensure that science and technology considerations are fully integrated into the foreign policy process. I am gratified that this conference deals with the policy dimensions of global challenges—not just the technical alternatives—because we are looking for ways for engineers to participate effectively in formulating policy, not just in presenting technical options. I do not overestimate how much I, an individual has-been chemist can do to influence U.S. foreign policy. However, the number one priority for my office is to establish the closest possible links between the foreign policy community and the men and women of science and engineering. I want to build a "superconducting bus-bar" between the National Academies and the State Department. I want to establish a seamless mechanism that ensures that the Department of State can readily draw on the best science and engineering talent and data in creating U.S. policy.

We have begun this process and it is working, but we have a long way to go. Even though there are excellent transmitters in the science and technology community, we will have to work hard to ensure that there are good receivers at the State Department. That will not be as easy as it seems. We face some serious challenges. First, we would like to be ahead of the curve, not limping along behind trying to catch up. In other words, we would like an early warning system that alerts policy makers to scientific and technological issues on the horizon that will become future policy problems. However, if the warnings arrive too early, they won't be heeded. The timing has to be right.

A second problem is that we rarely have all of the scientific data before a political decision has to be made. In addition, scientists and engineers often have different analyses and interpretations of the data that are available. That presents a huge dilemma for policy makers. Whom does one believe? For engineers to play a role in policy making, they must tell us where facts stop and opinions begin, and they must educate us about the reasoning behind the opinions. Engineers must also help us understand the risk-reward ratios for a given set of actions.

Finally, we must consider how the public views a given policy. Policies that do not have the support of the people will ultimately fail. Educating the public is an area in which the science and engineering community can do much more. On complex issues, it is not enough to tell the policy wonks at the State Department what they should do.

For instance, today in Europe political pressures have escalated the concept of "precaution" to such a high point that policy makers are demanding zero risk from genetically modified foods, environmental pollutants, and new energy sources. We all know nothing has zero risk—not even going out the door and crossing the street. And I assure you that trying to influence the elusive, often chaotic process of formulating our nation's foreign policy is not a zero-risk proposition. That is why we want you to help us. You must pay more attention to explaining your views and the technical bases for them in the popular media so the public can make rational, informed choices.

There is perhaps one more useful thought about risk. *Scientific American Presents* included some comments on extreme engineering in the winter 1999 issue that you might remember when you consider complex global systems. The article includes a sobering aerial photograph of Pripyat, an abandoned town near Chernobyl, a ghostly, ghastly gray under a layer of snow and lowering winter clouds. The caption reads: "Colossal accidents happen when overconfidence and complacency prevail." That's a useful reminder in any discipline—even foreign policy. The article also says that "engineers and managers of technology, being human, can come to believe in themselves and their creations beyond reasonable limits" (Petroski, 1999). Although this always leads to failure initially, once the failure is understood and the sting of tragedy is sufficiently remote, engineers usually pick up where they left off in pursuit of greater goals—which they then often attain.

Just one final point: in this new disorderly world, I believe we must change our definition of national security. There is more to ensuring our nation's security than the intelligence community and the military can provide. The front line of national security is still diplomacy, and our embassies and consulates abroad who are seeking solutions for political unrest, negotiating global treaties to protect or reclaim the environment, stimulating economic growth and development, helping ease the burdens of disease that can inhibit economic progress and lead to regional instability, and working in countless other ways to build and sustain peaceful, constructive relationships among nations.

But diplomats are not always successful. Diplomacy is a lot more than sipping tea and attending cocktail parties. It is the last line of defense before war. At this point in human history, war—despite its popularity among the despots and the desperate—on a global scale is no longer a viable option. Therefore, the Department of State is a vital instrument of national security.

Unfortunately, the department has not fared well in the annual budget wars up and down Pennsylvania Avenue. Since 1985, the total budget for international affairs has fallen some 34 percent in real terms. In the last five years the budget for State Department operations has dropped 17 percent. During the same period we have had to establish new embassies abroad as countries have emerged from ethnic conflicts or the collapse of authoritarian regimes. We have had to strengthen our buildings against expanding terrorist threats—and the State Department has had its casualties. The lack of resources has delayed the deployment of cutting-edge communications and computer technology for handling the flow of information vital to effective diplomacy. And, as the NRC report pointed out, the limited resources have constrained our ability to coordinate the international science and technology initiatives that are often a significant component of our diplomatic gestures toward other countries. Budgetary constraints have also limited the number of technically qualified people in the State Department who can be effective receivers of the advice and counsel of engineers in the policy-making process.

These are real issues, and they need real attention. This time, at least, William Safire was right—"it's the world, stupid."

REFERENCES

National Research Council. 1999. The Pervasive Role of Science, Technology, and Health in Foreign Policy. Washington, D.C.: National Academy Press.

Petroski, H. 1999. The hubris of extreme engineering. Scientific American Presents 10(4): 94–104.

Panel I

Understanding, Adapting, and Mitigating Climate Change through Engineering

As our understanding of the dynamics of climate change improves, we must initiate organized efforts either to influence or to adapt to these changes. This panel addresses the role of the engineering community in understanding and responding to changes in Earth systems.

Climate Systems Engineering

ROBERT M. WHITE

People have attempted to manipulate earth systems for thousands of years. Primitive engineering tended to focus on basics, such as shelter, water resources, and transportation, and little thought was given to the ancillary and frequently deleterious consequences of human activities. Addressing these consequences has now become the focus of attention.

Climate systems engineering, a subset of Earth systems engineering, is a multipurpose, multidisciplinary approach to monitoring, adapting to, and even mitigating the consequences of climate change. For much of history, climate change was regarded as an act of God over which people had no control. Recently, however, human activities have been acknowledged to be at least partly responsible for climate change, particularly global warming. Today, climate change is a topic of intense national and international interest because of its environmental, economic, and social consequences.

Modern weather and climate sciences are largely the result of advances in engineering and technology. Scientific weather forecasting became possible 150 years ago with the introduction of the telegraph, which provided a mechanism for transmitting weather information from remote areas to central locations where they could be analyzed. Since World War II, a host of other technologies has revealed some of the mysteries of climatic conditions. Radiosondes provide a view of the upper atmosphere; radar has transformed our understanding of the dynamics of precipitation and cloud systems; computers have enabled us to construct mathematical models of weather and climate. Very recently, space technology has provided imaging, sounding, and location capabilities making possible the global monitoring of weather and climate. All of these have transformed climate prediction from an art to a science.

Since the beginning of the Industrial Age, the increasing use of fossil fuels, deforestation, and emissions from other sources have dramatically increased atmospheric concentrations of greenhouse gases. Annual emissions of carbon dioxide (CO_2), for example, have increased from barely detectable levels 140 years ago to more than 6 billion metric tons per year (Figure 1). Observations at observatories around the world have shown that the increase has been essentially monotonic, except for seasonal fluctuations. The result of this 25-percent increase (from approximately 290 parts per million by volume [ppmv] in 1860 to about 360 ppmv in 1999) has resulted in a rise in global mean surface temperature (Figure 2). The temperature change is a matter of observational fact about which there is little dispute.

According to the Intergovernmental Panel on Climate Change (IPCC), mathematical climate models indicate that global surface temperatures will increase significantly by 2100 (IPCC, 2001). Most models project an increase of 1.5 to 4.5°C. The latest IPCC assessment estimates a temperature rise of 1.5 to 5.8°C, most likely about 2.5°C and an increase in global precipitation. Sea level is also rising, largely because of the thermal expansion of seawater; the rise is predicted to be approximately 0.5 meter.

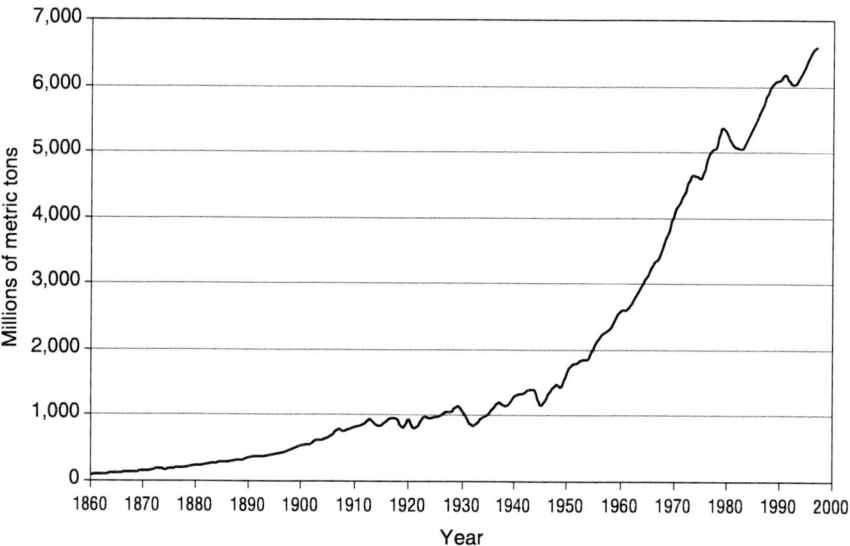

FIGURE 1 Global anthropogenic CO_2 emissions, 1860–1994. Source: Adapted from Marland et al., 2000.

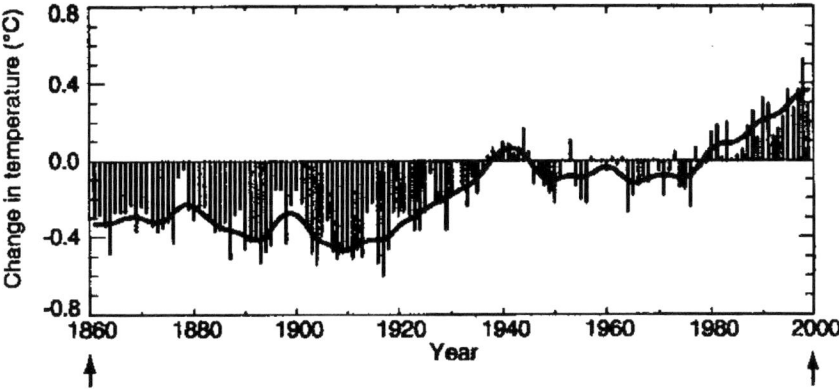

FIGURE 2   Global mean surface temperatures. Source: IPCC, 2001.

The regional distribution of climate change and its impact on agriculture, ecosystems, and water resources, as well as its effect on severe weather, such as hurricanes, are less certain. In a recent report on the impact of climate change on the United States, the National Assessment Synthesis Team (2000) focused on the consequences through 2100 for five sectors of the economy and 16 geographical regions. The analysis was based on two different climate models, one developed by scientists in Canada and the other by scientists in the United Kingdom. Both models project warming for the United States as a whole, but their regional projections differ significantly.

Despite the uncertainties, projections of future climate change present us with a serious dilemma. How should we balance the costs of the economic and social impacts of climate change with the costs of engineering and technology that could prevent those consequences? When and at what cost should we build dams and seawalls and stronger bridges? When should we turn to biotechnology, such as drought-resistant and heat-resistant strains of grain?

Almost 10 years ago, in the Framework Convention on Climate Change (FCCC), the international community agreed to try to "achieve stabilization of greenhouse gas concentrations in the atmosphere at a level that would prevent dangerous anthropogenic interference with the climate system" (United Nations, 1992). The convention addressed CO_2 and other greenhouse gases, including methane, ozone, and nitrous oxide. Because CO_2 dominates the greenhouse gas mixtures, however, the following discussion is focused on CO_2.

The convention did not define "dangerous," but any definition must include familiar, sometimes devastating, phenomena, such as threats to the food supply and water resources; rising sea level that can lead to inundations of islands and coastal areas; an increase in severe weather, such as hurricanes, floods, and droughts; changes in ecosystems; and adverse health effects, such as increased pulmonary and cardiovascular diseases. Mitigating global warming is one of the most important, difficult, and complex challenges ever faced. The prime causes of elevated global CO_2 concentrations are CO_2 emissions from the combustion of fossil fuels and deforestation (Table 1). Humanity's addiction to fossil fuels (coal, gas, and oil) as sources of energy is responsible for much of the rise. Attempts to solve the problem by decarbonizing the global energy supply were begun more than a century ago (Figure 3).

Since the climate convention was initialed in 1992, governments around the world have been looking for ways to control atmospheric greenhouse emissions without setting specific targets for atmospheric concentrations. The Conference of the Parties (COP), a group established to negotiate the details of the FCCC, has met six times to work out an agreement on international action. A meeting at The Hague in late 2000 ended in disagreement.

A protocol initialed by the COP in Kyoto, Japan, in 1997 would limit emissions of CO_2 and other greenhouse gases and assign emission targets to industrialized countries. Developing countries unwilling to commit to the protocol were given a pass (United Nations, 1997). The Kyoto agreement requires that the United States reduce greenhouse gas emissions to a level 7 percent below 1990 levels by 2010. Achieving this reduction would require reducing U.S. consumption of fossil fuels about 35 percent below the expected level in 2010. The reduction would require dramatic changes in energy production and energy use in this country.

The Kyoto signatories agreed that the sequestration of carbon in the biosphere, principally by trees, could be an ancillary approach for reducing greenhouse gas concentrations in the atmosphere. The United States proposed that it be permitted to use carbon sequestration by forest and agricultural lands and emissions trading with other countries to account for about 50 percent of the required reduction. This proposal was rejected by the European members of COP and was largely responsible for the collapse of the Hague conference. Alternative scenarios for meeting the Kyoto targets that focus more on non-CO_2 gases were proposed by Hansen et al. (2000).

TABLE 1 Principal Sources of Anthropogenic CO_2 Emissions

Source	Gigatons of carbon per year (GtC/yr)
Fossil fuel combustion	5.5 ± 0.5
Deforestation	1.6 ± 1.0
Total anthropogenic emissions	7.1 ± 1.1

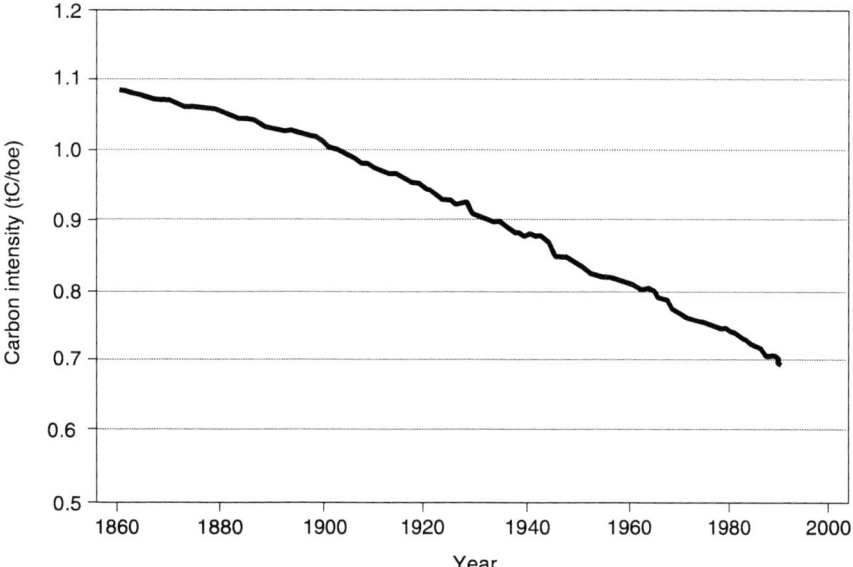

FIGURE 3 Decarbonization: carbon intensity of global energy consumption (tons of carbon per ton of oil equivalent energy [tC/toe]). Source: Nakicenovic, 1996.

Even if the Kyoto protocol were successful, there is general agreement that it would only be a first step that would have only a minimal effect on projected global warming. According to the IPCC, emissions targets in the protocol would reduce global average temperatures by an insignificant amount. Emissions reductions of 60 to 80 percent would be necessary to stabilize atmospheric CO_2 concentrations at their present levels. The agreement has also been fraught with political controversy because China, India, and other developing countries have been unwilling to restrict their emissions.

Anticipating the wrenching changes that will be required and aware that not all nations will be required to reduce emissions, the U.S. Senate voted unanimously against U.S. participation in the Kyoto protocol. In addition, in 2001 the Bush administration announced that it would not regulate CO_2 emissions from power plants and indicated that the United States would withdraw from the Kyoto protocol.

Minimizing, or even reversing, the adverse effects of climate change is an extremely complex challenge. The first step will be setting target levels of global greenhouse gas concentrations, a process that is rife with uncertainties and controversy. Setting targets will require knowledge of (or plausible estimates of) the consequences of specific limits. A commonly discussed target, greenhouse gas concentrations roughly double those of the era before the Industrial Revolution, would yield a concentration of about 550 ppmv, which many believe would prevent dangerous interference with the climate system. Engineers and technologists are faced with a mind-boggling array of options for achieving specific target concentrations of greenhouse gases. Options include reducing emissions of CO_2 from fixed and mobile sources, sequestering carbon dioxide, reducing the emissions of other greenhouse gases, and using geoengineering (e.g., injecting dust or other particulate matter into the stratosphere to reduce the amount of solar radiation reaching Earth) on a global scale (Box 1).

BOX 1
Options for Reducing Concentrations of Atmospheric Greenhouse Gases

Reductions in Emissions of CO_2 and Other Greenhouse Gases

- Increase the efficiency of both mobile and fixed sources of CO_2 (e.g., the Partnership for a New Generation of Vehicles [PNGV]).
- Increase the efficiency of electric power generation by changing power-station fuel sources from coal and oil to gas and by introducing turbines and distributed energy sources.
- Increase the use of renewable energy sources, such as wind power, photovoltaics, biomass, and hydropower.
- Increase the use of already-proven nuclear energy, a CO_2-emission-free energy source that is widely used for power production in France and other countries.
- Continue the development of new types of energy systems, such as fuel cells for use in automobiles and in fixed locations operating on hydrogen stripped from fossil hydrocarbons.

Carbon Sequestration

- Increase sequestration by growing trees and other plants, which consume carbon dioxide through photosynthesis. This approach can be enhanced through biotechnology by producing fast-growing trees. Carbon might also be sequestered in soil.
- Sequester carbon stripped from hydrocarbons by pumping it into deep geological structures, and use the hydrogen to power fuel cells.

Geoengineering

- Inject CO_2 into oceans at depths that allow the formation of CO_2 hydrates.
- Fertilize the oceans by adding iron or phosphorus to increase the production of algae, which would then sequester more carbon in the oceans.
- Disperse dust or inject SO_2 into the stratosphere to reduce sunlight and thereby lower global temperatures.

Even if new, low-carbon energy systems are developed, carbon is sequestered, and concentrations of other greenhouse gases can be reduced, much more will have to be done. We will have to anticipate the consequences of climate change on ecosystems, water resources, agriculture, health, and other human concerns. We will also have to consider the effects of new technologies on human health and ecosystems. In addition, these technologies will have to be culturally acceptable. These considerations cannot be afterthoughts; they must be included in the requirements and development of a technology. Answering these and many more questions will require extensive collaboration between scientists and engineers in many fields.

Because of the global nature of climate change, the issue has both domestic and international dimensions. In addition to participating in international negotiations on atmospheric concentrations of CO_2, we must reach out to engineering communities in other countries to enlist their help. The task before us is formidable. A wise course of action would be to take action to reduce emissions and sequester carbon now and, at the same time, invest in research and engineering to generate new technologies that will enable us to meet target concentrations in the future.

REFERENCES

Hansen, J., M. Sato, R. Ruedy, A. Lacis, and V. Oinas. 2000. Global warming in the 21st century: an alternative scenario. Proceedings of the National Academy of Sciences 95: 9875–9880.

IPCC (Intergovernmental Panel on Climate Change). 2001. Climate Change 2001: Synthesis Report: Third Assessment Report of the IPCC. New York, N.Y.: Cambridge University Press.

Marland, G.T., T. Boden, and R.J. Andres. 2000. Global CO_2 Emissions from Fossil-Fuel Burning, Cement Manufacture, and Gas Flaring: 1751–1997. Available online at *<http://cdiac.esd.ornl.gov/ftp/ndp030/global97.ems>* (April 16, 2001).

Nakicenovic, N. 1996. Freeing energy from carbon. Daedalus 125(3): 95–112.

National Assessment Synthesis Team. 2000. Climate Change Impacts on the United States: An Overview. Cambridge, U.K.: Cambridge University Press.

United Nations. 1992. United Nations Framework Convention on Climate Change. Available online at *<http://www.unfccc.de/resource/conv/index.html>* (March 21, 2001).

United Nations. 1997. United Nations Kyoto Protocol. Third Session of the Conference of the Parties to the Framework Convention on Climate Change in Kyoto, Japan, December 1997. Available online at *<http://www.unfccc.de/resource/protintr.html>* (March 22, 2001).

Improving Climate Assessment

ROBERT W. CORELL

How do we as a nation, particularly the engineering and technology communities, address the opportunities and challenges posed by changes in climate? Global climate change affects every sector of our economy and society, from the quality and availability of food and water to human health, coastal and ocean processes, forests, ecosystems, and energy demand and supply. In the last decade, a methodology has been designed to provide decision makers in the public and private arenas with an independent, multiscale assessment of the state of scientific knowledge. The policy community has always sought scientific advice and counsel through consultations with small, often informal groups of experts. The new "assessment model" provides a more structured approach that connects new discoveries and predictive capabilities with decision-making processes. This assessment process was designed to provide a consensus of the scientific and technical community of the problem and to estimate the consequences.

However, this assessment strategy has some weaknesses because it does not account for the hidden factors that, combined with climate change, produce an effect. These hidden factors include land use and land cover, shifts in population and demographics, socioeconomic trends, energy policies and practices, and available resources. Despite these weaknesses, however, this methodology has served us well. Recent reports by the International Panel on Climate Change (Watson et al., 1996), a companion assessment of climate change in the United States by the National Assessment Synthesis Team (NAST, 2001), and the *International Assessment of Ozone Depleting Compounds* by the World Meteorological Organization (1994) were all based on this model.

An alternative model, the so-called "vulnerability model," accounts directly for all of the factors contributing to an effect. The vulnerability model is based on a simple framework that integrates impacts with mitigation and adaptation strategies (Figure 1). We intuitively understand the concept of vulnerability. For example, if a health threat, such as a new infectious disease arises, the public health community rapidly provides drugs or therapies to address the threat. The smaller the difference between the impact (the potential for a disease) and the adaptation strategy (the effectiveness of the drug) the smaller the vulnerability to this health threat. Occasionally, a virulent, highly contagious disease like the Ebola virus arises. This virus acts so quickly that there is virtually nothing we can do to adapt. Hence, in this case, our vulnerability is very great.

In responding to climate change, the engineering and technological community can both mitigate the effects and adapt our capabilities. Mitigation strategies can reduce the impacts, thereby reducing vulnerabilities; vulnerabilities can be reduced further by improving our adaptative capacities. We need a new "calculus," an improved vulnerability model that connects these three elements (the climate change, mitigation strategies, and adaptations), both conceptually and analytically, that would enable us to address our vulnerabilities more holistically.

The vulnerability approach is based on an understanding that accelerating rates of change in the Earth's climate, air and water quality, and humankind's use of land, natural resources, and ecosystems affect the well-being of societies and

---

Vulnerability (V_n) can be seen as the difference between the cumulative impacts (I_n) from multiple stressors and mitigation strategies (m_n) and the adaptive capacity (AC_n) or:

$$V_n = I_n \text{ (where } I_n \text{ is a function of } m_n) - AC_n$$

Note: First, a number of vulnerabilities, impacts, and adaptive capacities are denoted as I_n in this description. Second, the elements of the equation all vary with time and on several spatial scales. Finally, research is under way to transform this simple arithmetic form into a more powerful analytical multidimensional matrix vector calculus form.

FIGURE 1 The vulnerability model.

the overall quality of life on our planet. Addressing this complex reality will require better models and methodologies. Many governments, businesses and industries, communities, and individuals are looking for ways to understand the ultimate consequences of these changes and develop strategies for coping with them. As we become more aware of the complexity of these changes, our need for a better assessment methodology increases. As vulnerability analysis evolves, other factors, such as social inequities, poor health, inadequate environmental services, and lack of access to state services, infrastructure, and other essential resources, that contribute to the effects of climate change are being identified and integrated.

This conceptual framework has been improved by the development of powerful quantitative methods and mathematical models that have enabled the scientific and engineering community to analyze how causes and impacts overlap, cluster, aggregate, and interact. Qualitative variables that complement the quantitative elements and methodologies are central to the model. Vulnerability analyses can also take into account the interests of individuals, groups, sectors, and nations, as well as the systems in which they are embedded, to determine the vulnerability of a specific community or environmental system to multiple social stresses, environmental stresses, and climate changes.

The report by the National Assessment Synthesis Team included 10 key findings suggesting opportunities and challenges for the engineering and technological communities (NAST, 2000).

1. *Increased Warming.* Assuming continued growth in world greenhouse gas emissions, the primary climate models used in this Assessment project are that temperatures in the U.S. will rise 5° to 9°F (3° to 5°C) on average in the next 100 years. A wider range of outcomes is possible.
2. *Differing Regional Impacts.* Climate change will vary widely across the United States. Temperature increases will vary somewhat from one region to the next. Heavy and extreme precipitation events are likely to become more frequent, yet some regions will get drier. The potential impacts of climate change will also vary widely across the nation.
3. *Vulnerable Ecosystems.* Many ecosystems are highly vulnerable to the projected rate and magnitude of climate change. A few, such as alpine meadows in the Rocky Mountains and some barrier islands, are likely to disappear entirely in some areas. Others, such as forests of the Southeast, are likely to experience major species shifts or break up into a mosaic of grasslands, woodlands, and forests. The goods and services lost through the disappearance or fragmentation of certain ecosystems are likely to be costly or impossible to replace.
4. *Widespread Water Concerns.* Water is an issue in every region, but the nature of the vulnerabilities varies. Drought is an important concern in

every region. Floods and water quality are concerns in many regions. Snowpack changes are especially important in the West, Pacific Northwest, and Alaska.

5. *Secure Food Supply.* At the national level, the agriculture sector is likely to be able to adapt to climate change. Overall, U.S. crop productivity is very likely to increase over the next few decades, but the gains will not be uniform across the nation. Falling prices and competitive pressures are very likely to stress some farmers, while benefiting consumers.

6. *Near-Term Increase in Forest Growth.* Forest productivity is likely to increase over the next several decades in some areas as trees respond to higher carbon dioxide levels. Over the longer term, changes in larger-scale processes such as fire, insects, droughts, and disease will possibly decrease forest productivity. In addition, climate change is likely to cause long-term shifts in forest species, such as sugar maples moving north out of the United States.

7. *Increased Damage in Coastal and Permafrost Areas.* Climate change and the resulting rise in sea level are likely to exacerbate threats to buildings, roads, power lines, and other infrastructure in climatically sensitive places. For example, infrastructure damage is related to permafrost melting in Alaska, and to sea-level rise and storm surge in low-lying coastal areas.

8. *Adaptation Determines Health Outcomes.* A range of negative health impacts is possible from climate change, but adaptation is likely to help protect much of the U.S. population. Maintaining our nation's public health and community infrastructure, from water treatment systems to emergency shelters, will be important for minimizing the impacts of water-borne diseases, heat stress, air pollution, extreme weather events, and diseases transmitted by insects, ticks, and rodents.

9. *Other Stresses Magnified by Climate Change.* Climate change will very likely magnify the cumulative impacts of other stresses, such as air and water pollution and habitat destruction due to human development patterns. For some systems, such as coral reefs, the combined effects of climate change, and other stresses are very likely to exceed a critical threshold, bringing large, possibly irreversible impacts.

10. *Uncertainties Remain and Surprises Are Expected.* Significant uncertainties remain in the science underlying regional climate changes and their impacts. Further research would improve understanding and our ability to project societal and ecosystem impacts, and provide the public with additional useful information about options for adaptation. However, it is likely that some aspects and impacts of climate change will be totally unanticipated as complex systems respond to ongoing climate change in unforeseeable ways.

In some geographical areas, such as coastal regions, the report concluded that engineering strategies could help communities adapt to the multiple effects of climate change in coming decades. The report also concluded that the adaptive capacity of agricultural systems could be increased through biotechnology, which could provide drought-resistant seeds, and through adaptive planting strategies. For rare ecosystems, such as coral reefs, which have very little capacity to adapt, mitigation strategies are a more likely way to reduce vulnerabilities. A number of these ecosystems are expected to migrate northward as temperatures rise and precipitation increases.

The fundamental purpose of all of these analyses is to improve our understanding of how the climate system works. Over a period of decades the science and technology community has made substantial progress in modeling the climate system and projecting changes. With the advent of Earth systems engineering and improved vulnerability models, the engineering and technology communities can contribute to solutions. With continued improvements in Earth systems engineering, we should be able to adapt to and mitigate climate changes in the coming decades.

REFERENCES

NAST (National Assessment Synthesis Team). 2000. Climate Change Impacts on the United States: An Overview. Cambridge, U.K.: Cambridge University Press.

Watson, R.T., M.C. Zinyowera, and R.H. Moss, eds. 1996. Climate Change 1995: Impacts, Adaptations, and Mitigation of Climate Change. Cambridge, U.K.: Cambridge University Press.

World Meteorological Organization. 1994. Scientific Assessment of Stratospheric Ozone, 1994. Report No. 37. Geneva, Switzerland: World Meteorological Organization.

How Camest Thou in This Pickle?

JERRY M. MELILLO

Addressing the problem of increasing concentrations of carbon dioxide in the atmosphere brings to mind a simple question from Shakespeare's *The Tempest*, "How camest thou in this pickle?" The answer is anything but simple. The developed world has so far been primarily responsible for disrupting the global carbon cycle. However, in the future parts of the developing world are likely to become the dominant emitters of carbon dioxide (CO_2) through the burning of fossil fuels. I believe developed nations should lead by example by reducing their CO_2 emissions and promoting sensible programs to sequester carbon. I also believe developed countries should share carbon-efficient energy-generating technologies with developing countries as rapidly as possible. After all, a mole of CO_2 contributes to the greenhouse effect regardless of its source.

In attempting to address carbon cycle problems through engineering, we must anticipate that the interventions may have negative environmental consequences. We are dealing here with complex nonlinear ecosystems with yet-to-be-defined limits and thresholds, and we must move forward slowly. In addition to the global carbon cycle, humans have disrupted other major biogeochemical cycles, including those of nitrogen, sulfur, and phosphorus. Addressing these problems will require wise engineering combined with a deep understanding of ecological, economic, and social systems. Actions taken by and in the developing world will be crucial.

Earth's climate is a function of complex interactions among the sun, atmosphere, oceans, land, and living things. Several gases in the atmosphere, including CO_2, absorb heat radiated from the Earth's surface and create the "greenhouse effect," a natural feature of our climate system. Humans have been changing the composition of the atmosphere by burning fossil fuels and clearing forests for agriculture and other uses for the past 1,000 years. But until about 100 years ago these activities had a minor effect on the global carbon cycle and the climate system (NAST, 2000). Since the late 1800s, increasing emissions associated with human actions have been responsible for a 30 percent increase in the concentration of atmospheric CO_2. Many aspects of climate, including warming, have also occurred (Figure 1).

From 1950 to 1995 the developed world accounted for about three-quarters of total CO_2 emissions associated with the burning of fossil fuels. In 1995, for example, 73 percent of total CO_2 emissions from human activities came from developed countries (OSTP, 1995). The United States was the largest single source, accounting for 22 percent of the total, with carbon emissions per person exceeding 5 tons per year. Elsewhere in the developed world, Western Europe accounted for 17 percent, Eastern Europe and the former Soviet Union for 27 percent, and Asia for 7 percent. China was the largest single source among developing countries, accounting for 11 percent of the total, with carbon emissions per person about one-tenth those of the United States (Figure 2).

In the next few decades as much as 90 percent of the world's population growth is expected to occur in developing countries (Figure 3), some of which will concurrently undergo rapid economic growth. Per capita energy use in developing countries, which is now only one-tenth to one-twentieth of U.S. energy use, will also rise. If present trends continue, developing countries will account for more than half of total global CO_2 emissions by 2035. China—today the second-largest source of CO_2 emissions—will become the largest emitter sometime between 2010 and 2015.

As scientists currently understand the global carbon cycle, natural carbon sinks in the ocean and on land eventually absorb between one-half and two-thirds of emissions from human activities. The rest of the emitted carbon remains in the atmosphere. Therefore, to reduce the rate at which CO_2 accumulates in the atmosphere, we must either reduce emissions or increase carbon uptake by the land and the oceans.

Both developed and developing countries must eventually be involved in managing the emissions side of the global carbon cycle. Developed nations should lead the way because they are responsible for most of the CO_2 that has accumulated in the atmosphere since the late 1800s. Developed countries must also share carbon-efficient, energy-generating technologies with the developing world as soon as possible because developing nations with rapidly growing economies, such as China, are making capital investments now in power plants with lifetimes of at least several decades. It is in everyone's interest that these power plants be as carbon efficient as possible.

Both land and ocean ecosystems have the capacity to store or sequester carbon. On land vegetation currently stores about 550 billion metric tons of carbon and soil stores another 1,500 billion metric tons. The ocean, which is a

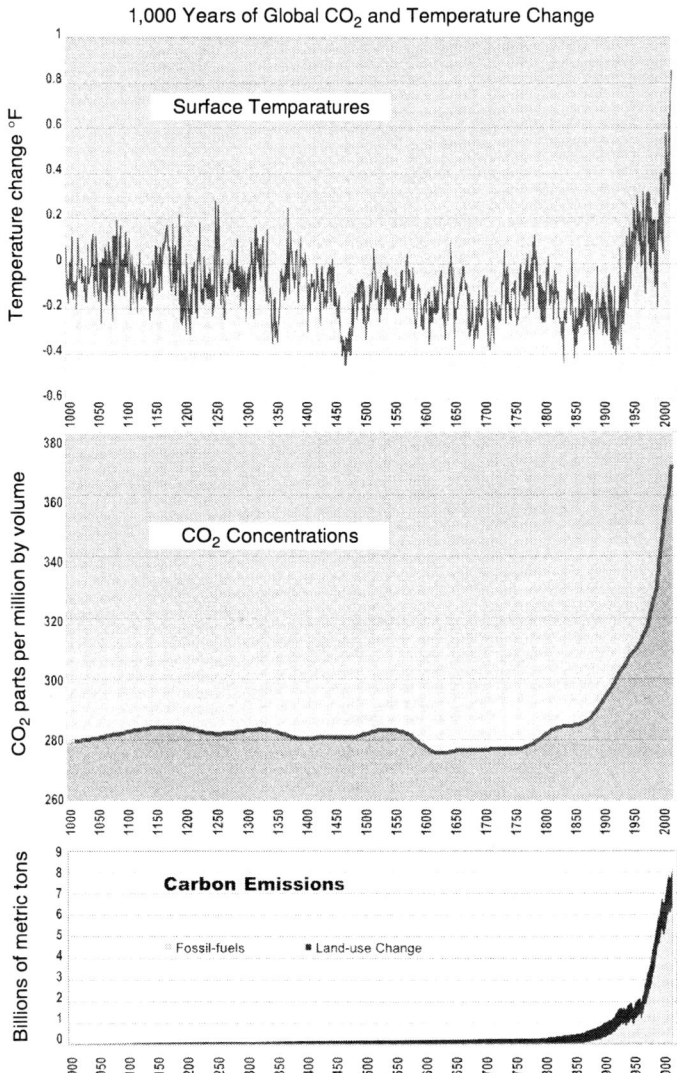

FIGURE 1 Records of surface temperatures, CO_2 concentrations, and carbon emissions in the Northern Hemisphere. **Surface Temperatures:** Reconstruction of annual average surface air temperatures derived from historical records, tree rings, and corals (until about 1900) and direct measurements of air temperatures (after about 1900). **CO_2 Concentrations:** Derived from measurements of CO_2 concentration in air bubbles in the layered ice cores drilled in Antarctica (for period before 1957) and from atmospheric measurements (since 1957). **Carbon Emissions:** Reconstruction of past emissions of CO_2 from land clearing and fossil fuel combustion since about 1750 (and linearly projected back to zero in 1000). Source: NAST, 2000.

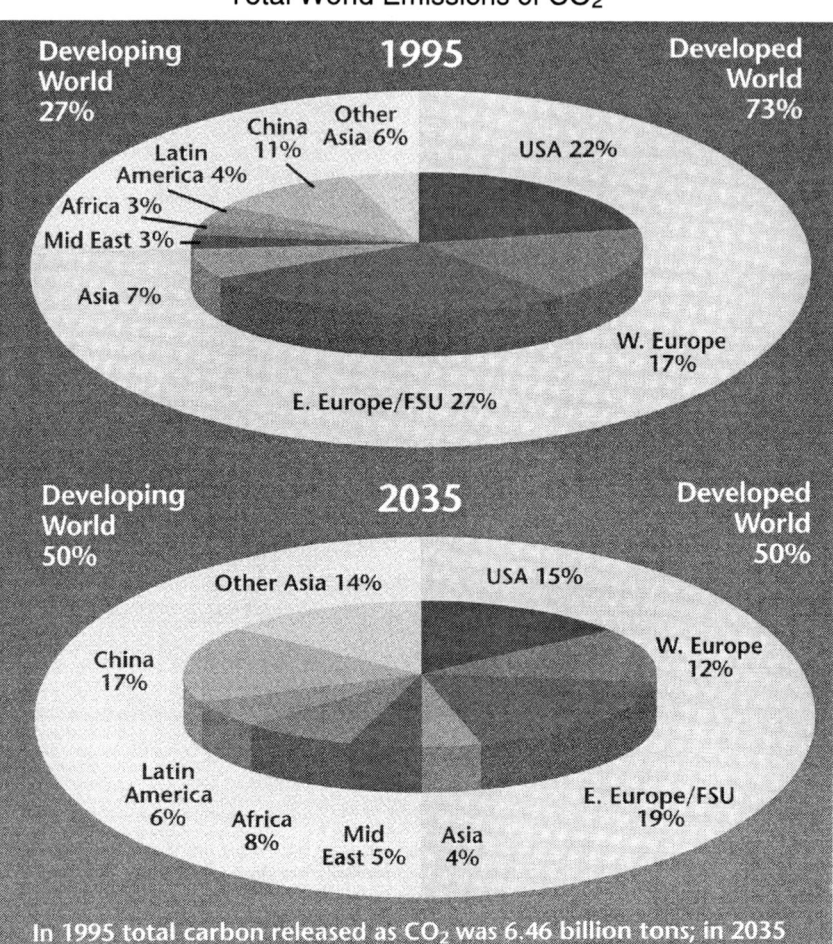

FIGURE 2 Relative distribution of total world CO_2 emissions associated with the burning of fossil fuels (estimated for 1995 and projected for 2035). Source: OSTP, 1995.

much larger carbon storehouse, contains about 38,000 billion metric tons of carbon, most of it in deep waters (Schlesinger, 1997).

In the past decade scientists and engineers have been exploring ways to increase carbon storage in both land and ocean ecosystems. Schemes for sequestering more carbon on land include reforestation (Birdsey and Heath, 1993) and the elimination of traditional tillage practices in managing agricultural soils (Lal et al., 1998). Interestingly, many of the proposed carbon sequestration schemes for land ecosystems would also have other benefits. For example, reforestation of hillsides would protect soil against erosion from heavy rains, and the buildup of organic matter in agricultural soils would increase their capacity to retain nutrients and water.

Schemes for increasing carbon storage in ocean ecosystems are not as well developed. One proposal that has received considerable attention would entail manipulating the biological component of the Southern Ocean in an attempt to store carbon in the deep ocean for centuries (Abraham et al., 2000; Boyd et al., 2000; Chisholm et al., 2000; Watson et al., 2000). This scheme is based on the observation that in the Southern Ocean a lack of available iron in sunlit surface waters limits the growth of phytoplankton—microscopic ocean plants—that form the basis of the marine food web. Using sunlight and dissolved nutrients, phytoplankton convert CO_2 to organically bound carbon. Animals eat the tiny marine plants, and microorganisms, such as bacteria, then decompose both plant and animal wastes. As the organic carbon passes through the marine food web, most of it is converted back to CO_2 and escapes into the atmosphere. A small amount, however, is transported to the deep ocean, where it, too, is eventually converted back to CO_2 but remains for about 1,000 years. The rate at which the CO_2 is "pumped" to the deep ocean is largely related to the composition of phytoplankton

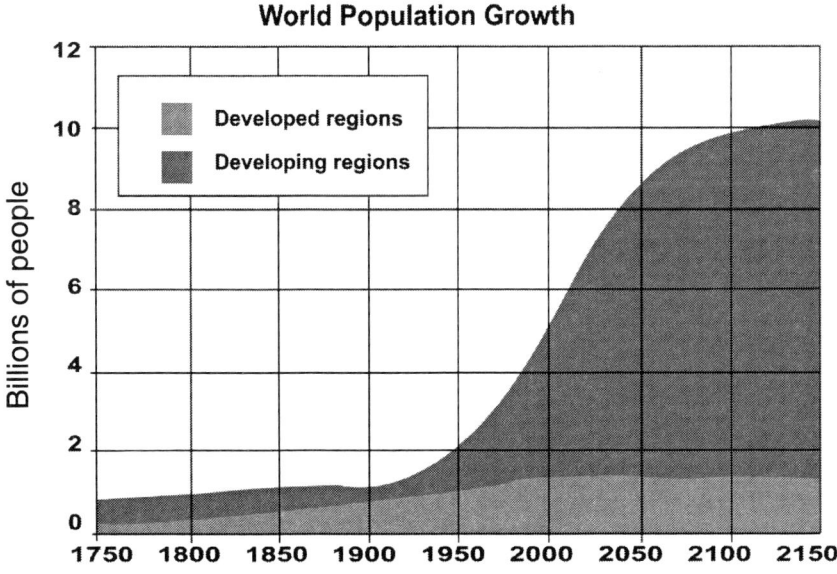

FIGURE 3   Population growth in the developed and developing world from 1750 to 2150.

species and their productivity, both of which are controlled by the availability of iron in the Southern Ocean. The idea is to add iron to this region of the ocean to increase carbon sequestration in deep waters.

This concept was tested in February 1999 when 8,663 kilograms of an iron compound were added to a circular patch of ocean 8 kilometers in diameter, located 2,000 kilometers south-southwest of Hobart, Tasmania (Boyd et al., 2000). As expected, iron fertilization led to a dramatic phytoplankton bloom and a shift in this community from small-celled to large-celled species, primarily diatoms. In a report on the results in *Nature*, Watson et al. (2000) conclude that "the experiment confirms that modest sequestration of atmospheric CO_2 by artificial additions of iron to the Southern Ocean is in principle possible, although the period and geographical extent over which sequestration would be effective remain poorly known."

Even if iron fertilization of the Southern Ocean resulted in a modest increase in the sequestration of atmospheric CO_2, this increase could come at a high price. In a critique of this scheme, S.W. Chisholm (2000), an MIT-based marine biologist, has argued that it would threaten ocean ecosystems by changing the structure of the marine food web. She reasoned that the iron-fertilization scheme could also produce many unintended side effects, such as deoxygenation of the deep ocean and the generation of greenhouse gases that are more potent than CO_2.

Other elements in the environment contribute to climate change. The increasing demand for food has led to the clearing of forests for cropland and pastures and the recent addition of large quantities of nitrogen and phosphorus as fertilizer. Today the fixation of nitrogen associated with production of food and energy is greater than natural nitrogen fixation in terrestrial ecosystems. Our growing demand for energy has also resulted in the burning of fossil fuels, wood, and other forms of biomass. These activities are major sources, not only carbon but also of sulfur and nitrogen in the atmosphere, where they affect the climate system and the chemistry of precipitation. The unprecedented mobilization of nitrogen, sulfur, and phosphorus has led to a range of environmental consequences at the local, regional, and global scales.

In the past century, most of the nitrogen, sulfur, and phosphorus mobilization has occurred in North America and Europe (Galloway et al., 1998). In the last few decades, however, Asia, which has more than half of the world population and many of the most rapidly growing economies, has substantially affected the global cycles of nitrogen, sulfur, and phosphorus (Figure 4). Estimates show that Asia today is affecting mobilization of these nutrients almost as much as North America and Europe combined. At the start of the new millennium, Asia accounted for 40 percent of nitrogen mobilization, 35 percent of sulfur mobilization, and 35 percent of phosphorus mobilization. Equally important is that mobilization in Asia is increasing rapidly. During the 1980s, for example, nitrogen mobilization in Asia from fertilizer alone doubled from 20 to 40 tetragrams of nitrogen per year (Tg N/year).

29

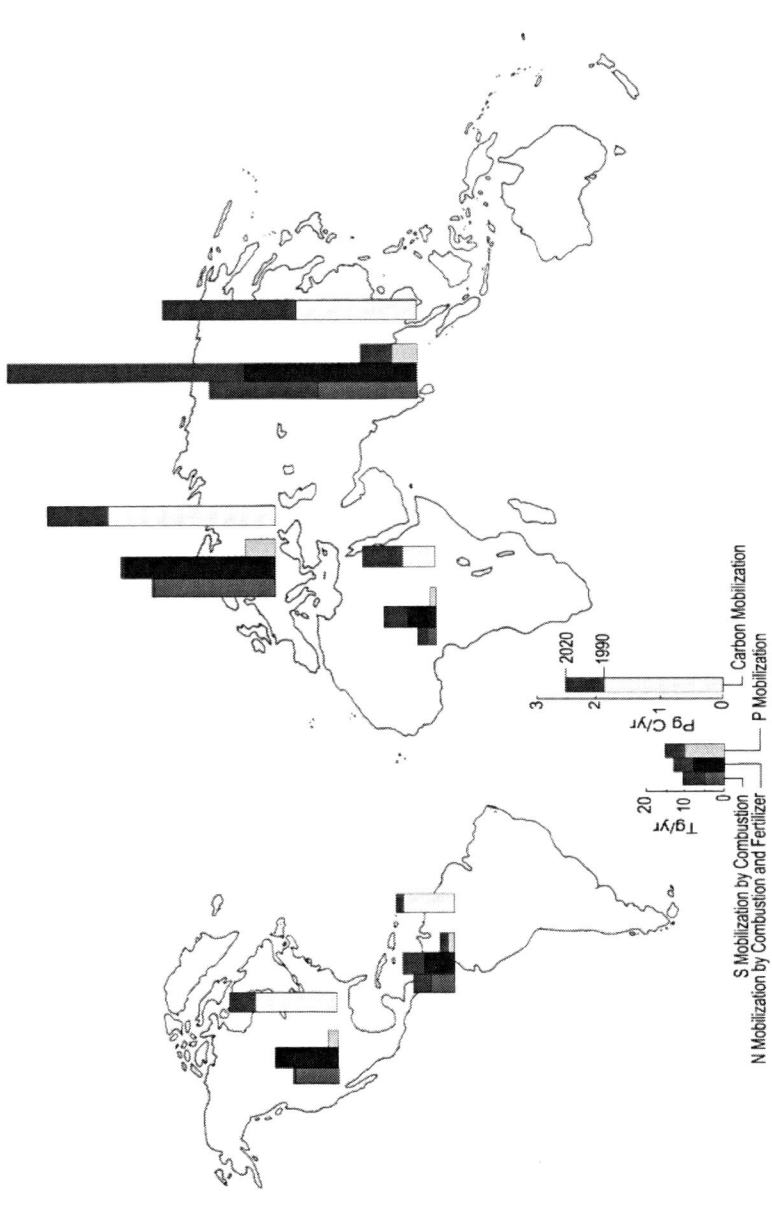

FIGURE 4 The mobilization of carbon, nitrogen, sulfur, and phosphorus in 1990 and 2020, by region. Source: Galloway et al., 1998.

Mobilized carbon, nitrogen, sulfur, and phosphorus ultimately accumulate in the Earth system's major reservoirs—the atmosphere, land, freshwaters, and oceans. The site and the magnitude of accumulation determine the environmental consequences. Nitrogen loading of the atmosphere as nitrous oxide contributes to the greenhouse effect and global climate change. Nitrogen loading can also lead to increases in tropospheric ozone and decreases in stratospheric ozone. Elevated levels of tropospheric ozone can cause human health problems and reduce crop production. Decreased levels of stratospheric ozone can mean that more ultraviolet radiation reaches the Earth's surface, causing human health problems. Loading of land ecosystems with nitrogen can acidify ecosystems. Loading of the atmosphere with sulfur increases its turbidity and acidity, which in turn can affect the radiation balance and acidify poorly buffered land and freshwater ecosystems. Finally, loading freshwater ecosystems with phosphorus can lead to a chain of events that includes increases in aquatic plant productivity, reductions in oxygen levels in the water column, and ultimately reductions in habitat quality for aquatic animals, including fish.

The rapid increase in carbon, nitrogen, sulfur, and phosphorus mobilization observed by Asia is expected to continue. By 2020 nitrogen mobilization is likely to double—from 45 Tg/year to 100 Tg/year. Asia is likely to consume about 50 percent of the phosphorus fertilizer used worldwide, compared with 35 percent in 1990. The region will also account for about half of all sulfur emissions to the atmosphere, compared with about 35 percent in 1990.

Continued increases in the rates at which nitrogen, sulfur, and phosphorus are mobilized will make managing our planet and sustaining and enhancing the quality of life at regional and local scales even more difficult. Developed countries must share their knowledge of the causes of atmospheric change and their technologies for enhancing our quality of life and reducing the adverse effects of disruptions to life-sustaining cycles of key elements.

Managing the Earth and its life-support systems will require a partnership among engineers, scientists, and policy makers. Together, we must develop and pursue an adaptive strategy in which new knowledge based on sound science and engineering is used to benefit humankind.

REFERENCES

Abraham, E.R., C.S. Law, P.W. Boyd, S.J. Lavender, M.T. Maldonado, A.R. Bowie. 2000. Importance of stirring in the development of an iron-fertilized phytoplankton bloom. Nature 407(6805): 727–730.

Birdsey, R.A., and L.S. Heath. 1993. Carbon Sequestration Impacts of Alternative Forestry Scenarios. Washington, D.C.: U.S. Department of Agriculture.

Boyd, P.W., A.J. Watson, C.S. Law, E.R. Abraham, T. Trull, R. Murdoch, D.C.E. Bakker, A.R. Bowie, K.O. Buesseler, H. Chang, M. Charette, P. Croot, K. Downing, R. Frew, M. Gall, M. Hadfield, J. Hall, M. Harvey, G. Jameson, J. LaRoche, M. Liddicoat, R. Ling, M. T. Maldonado, R. M. McKay, S. Nodder, S. Pickmere, R. Pridmore, S. Rintoul, K. Safi, R. Strzepek, P. Sutton, K. Tanneberger, S. Turner, A. Waite, and J. Zeldis. 2000. A mesoscale phytoplankton bloom in the polar Southern Ocean stimulated by iron fertilization. Nature 407(6805): 695–702.

Chisholm, S.W. 2000. Oceanography: stirring times in the Southern Ocean. Nature 407(6805): 685–687.

Galloway, J.N., D.S. Ojima, and J.M. Melillo. 1998. Asian change in the context of global climate change: an overview. Pp. 1–17 in Asian Change in the Context of Global Climate Change: Impact of Natural and Anthropogenic Changes in Asia on Global Biogeochemical Cycles, J.N. Galloway and J.M. Melillo, eds. Cambridge, U.K.: Cambridge University Press.

Lal, R., J.M. Kimble, R.F. Follett, and C.V. Cole. 1998. The Potential of U.S. Cropland to Sequester Carbon and Mitigate the Greenhouse Effect. Ann Arbor: Sleeping Bear Press.

NAST (National Assessment Synthesis Team). 2000. Climate Change Impacts on the United States: The Potential Consequences of Climate Variability and Change. Washington, D.C.: U.S. Global Change Research Program.

OSTP (Office of Science and Technology Policy). 1995. Climate Change: State of Knowledge. Washington, D.C.: Executive Office of the President.

Schlesinger, W.H. 1997. Biogeochemistry: An Analysis of Global Change. New York: Academic Press.

Watson, A.J., D.C.E. Bakker, A.J. Ridgwell, P.W. Boyd, and C.S. Law. 2000. Effect of iron supply on Southern Ocean CO_2 uptake and implications for glacial atmospheric CO_2. Nature 407(6805): 730–733.

Panel II

Using Biotechnology for the Benefit of Humanity

Advances in science and engineering are creating powerful tools for harnessing biological activity for human use, but these tools entail risks—some real and some perceived. This panel discusses the informed, responsible use of emerging biotechnologies to address global problems.

Genetically Modified Organisms
An Ancient Practice on the Cusp

MAXINE F. SINGER

Scientists and engineers are trained to adjust to change. During my own undergraduate and graduate studies about 50 years ago, eminent biologists were still choosing up sides as to whether proteins or nucleic acids carried genetic information in chromosomes. Look where we are now!

But revolutionary change doesn't go down easily outside of the technical community. As Paul Ehrlich, the distinguished environmentalist, recently pointed out, "A major contemporary human problem is that the rate of cultural evolution in science and technology has been extraordinarily high in contrast with the snail's pace of change in the social attitudes and political institutions that might channel the uses of technology in more beneficial directions" (Ehrlich, 2000). The different rates of change is a setup for problems. Serious gaps can develop between engineers' and scientists' ideas about the natural world and those that are current in mainstream society.

A case in point concerns genetically modified plants. For much of the nineteenth century a significant part of what we now call biology was called natural history. Tramping around the countryside looking for new species of beetles, fossils, or plants was considered a charming and harmless pursuit by the wealthy leisured class. Then, around the middle of the century, three great discoveries signaled a new kind of biology. One was the formulation of cell theory—the concept that all organisms are composed of one or more living cells. The second was Mendel's elaboration of the laws of inheritance. The third was Darwin's concept of evolution and the origin of species. Darwin, of course, opened a huge gap between science and the public that still haunts us. By the end of the twentieth century these three paths had converged into one biology—a science that is extraordinarily sophisticated and productive, although somewhat less charming and less acceptable to some.

As originally conceived, genes, though real enough, were thought to have no substance; they were considered abstractions. Mendel showed that genes are discrete bits of information passed from parents to offspring. One gene dictated the color of peas, another one whether they were smooth or wrinkled, and so on. Every organism, he realized, had two genes for each discrete trait, such as pea color—one from its maternal parent and one from its paternal parent. Most important, Mendel learned that any particular gene—for example, the gene responsible for the color of a pea—could occur in different forms. Depending on the two forms that were present in an individual plant, the peas would be green or yellow. These different versions of genes are responsible for variations within a species, including the variation we see if we look around a room at different faces. Geneticists call the individual variants of a gene an "allele," such as the green allele and the yellow allele for pea color.

In most populations of organisms there are many alleles for a given gene, not just two. New alleles, known as mutations, can arise. Although the word mutant has a negative connotation, not every mutant allele is bad news. Some alleles give organisms an advantage over their cousins in a particular environment. The advantaged organisms reproduce more efficiently than their cousins, thus explaining Darwin's idea of natural selection. Breeders of animals and plants practice unnatural selection, relying on human intervention to ensure the efficient reproduction of a selected organism.

The earliest plant breeders, probably starting about 10,000 years ago, made use of allelic variations, although they were ignorant of the underlying mechanisms. They observed new, rare alleles in fields and, when they noticed a novel property that was advantageous, they bred it into standard varieties. Wild potatoes, for example, contain high levels of alkaloid toxins. At least 4,000 years ago central Andean populations began selecting and breeding potatoes, presumably with alleles that reduced the poison.

Today, of course, we know that genes are made of segments of DNA. Alleles can differ from one another in the sequence of the four DNA bases that constitute their genetic code. Some alleles have more draconian changes—large segments of DNA or even most of a gene may be lost. Other alleles differ not in the gene coding sequence but in the surrounding DNA sequences that regulate the level at which the gene operates, or even whether it operates at all under particular conditions.

This sort of modulation of gene activity, which biologists call gene expression, underlies one of the alleles that differentiates modern maize, or corn, from teosinte, corn's wild ancestor of the same species (*Zea mays*), which is indigenous to central Mexico. Very few wild plants are closely related to corn, and none of them—not even teosinte—looks very much like the corn we know. Teosinte is a bushy plant with many tassels (the organ that produces pollen) and many seed-bearing stalks. The stalks are an inch or two long and have two rows of tiny seeds, each of which is covered with a very hard case. Unlike corn, these seed stalks have no green casing, or husk. The seeds eventually fall to the ground, sowing next year's plants and providing food for birds, which also disperse the seeds.

Corn could never have arisen or been propagated by natural processes because the seeds, or kernels, are tightly attached to the cob and cannot disperse. Corn plants cannot propagate themselves without human intervention. About 5,000 years ago, Central American plant breeders began selecting and growing teosinte variants because they were advantageous. The fundamental differences between teosinte and the corn we know are accounted for by variant alleles in five genes. At least one of those fundamental changes—the one that makes corn grow as a single straight stalk rather than a bush like teosinte—reflects a change in a regulation of gene activity rather than in a coding segment of a gene. This is an extraordinary case of human engineering of a natural system.

Our modern diets are composed almost entirely of genetically modified organisms (GMOs). If that history were better understood, the current public debate about GMOs might have a different focus. Today, the term GMO is commonly used to refer to plants that have been modified by modern molecular techniques, and I will use it that way. Few people understand the continuum between ancient and modern methods.

Modern molecular techniques emerged about 30 years ago when biologists learned how to manipulate genes precisely through techniques variously called recombinant DNA and cloning. These techniques enabled researchers to make direct changes in DNA structure to accomplish a predetermined purpose. Rather than waiting for the chance emergence of a desirable allele and then breeding it into a variety of plant, biologists can now design alleles to meet their needs. The methods used in plants are essentially the same as those used to understand and develop treatments for human diseases. All of them generally entail changing only a small number—from one to several thousand—of the billions of base pairs in an organism's genome.

The new methods can yield all of the allelic changes that occur spontaneously, and in much less time than the five to ten years normally associated with traditional methods. Biologists can also introduce new genes into plants as they have done with the interbreeding of two species (such as crossing a pomelo and an orange to produce a grapefruit). In traditional interbreeding, successes are rare, and there is a significant probability that traits undesirable in terms of the environment or food safety will remain. In contrast, the new techniques are rapid, because they allow biologists to introduce a single change to a single gene, and the probability of introducing undesirable properties is much lower.

What kinds of genes or alleles are introduced? The possibilities include genes from varieties of the same species, genes from related species, and genes from totally unrelated species, including genes from bacteria and animals. This, of course, is very different from traditional breeding. The apparent strangeness of this idea, for example, putting a fish gene into a strawberry plant to protect the plant from frost, has elicited a great deal of discussion and misunderstanding. It's important, therefore, to consider exactly what we mean when we say that we're putting a bacterial gene or a fish gene into a plant.

Biologists first identify the appropriate gene—a segment of DNA—and then isolate it from the rest of the DNA of the source organism using the technique known as cloning. Usually this means allowing bacteria to reproduce the DNA segment and then chemically isolating it. Sometimes biologists can introduce the isolated DNA directly into a plant, but often they modify it first to make it more suitable for its new location. For example, the DNA code words might be changed to enable the gene to work more efficiently in its new plant host.

Biologists then introduce the gene into the new plant, sometimes by shooting it in and sometimes by transferring it on the DNA of a special bacterium that, in nature, transfers its own DNA into plants. The original gene may have come from a fish, but it has been modified and amplified in many different bacterial cells before it is inserted into the plant's genome. At that point it is a pure, definite chemical structure, a piece of DNA. Is it still a fish gene? That, I believe, is a philosophical question, not a scientific one.

Like all complicated problems, the question of whether genetically modified plants will be safe for human health and the environment has no simple yes or no answer. Even assuming they are safe, we can't even say whether they are desirable. Opinions are sure to differ, depending on who is answering the question. However, the issues raised are no different from those posed by new plant varieties produced by traditional breeding. The questions aren't focused on the process used to produce the plants but on the nature of the modified plant. Each type of modified plant must be assessed on its own merits in relation to its use and the environment in which it will be grown.

To do this, we must focus on several different classes of concern. In the case of food, we're interested in the safety of the engineered plants for human and animal consumption. The environmental effects of all modified plants, both positive and negative, must be weighed. The resolution of environmental issues depends on scientific information that may or may not be readily available. And some concerns are only partly answerable by science. These include economic and humanitarian concerns, such as the limited ability of poor people, largely in Africa and Asia, to gain access to new plant varieties.

Consider, for example, Bt corn and Bt cotton, which have been engineered to resist certain insect pests by enabling the plants to produce an insecticidal protein within their own cells. Corn and cotton have also been bred for insect resistance through traditional breeding methods based on natural plant alleles that make plants resistant to some insects. Insects can also be controlled by chemical spraying of fields. So the question is one of balance: which method is preferable—spraying, traditional breeding, genetic engineering, or some combination? For genetically engineered and traditionally bred food plants alike, we must ask whether newly introduced changes yield a protein that's allergenic or toxic to humans and animals. Has the amount of some toxic component in the normal plant been increased? If biologists used an antibiotic-resistant marker gene for convenient manipulation, as they often do, we might be concerned that the effectiveness of an important drug might be compromised.

Five years ago U.S. farmers began planting Bt corn and cotton. These crops contain copies of genes coding for proteins that are toxic to a major corn pest, the European corn borer, and other pests that destroy cotton. The genes were copied from a bacterium called *Bacillus thuringiensis*, or Bt for short. As of summer 1999, more than 30 percent of the corn and 27 percent of the cotton planted in the United States contained Bt, a total of 30 million acres (Vorman, 1999). The underlying purpose is to reduce the 30 to 40 percent of the crop that is lost to pests each year worldwide. Organic farmers have used *Bacillus thuringiensis* itself by the ton for more than 40 years to control insect pests, so biologists had good reason to think that the Bt toxin would be harmless. A lot of the engineered corn is fed to animals or goes into products like corn oil that we've all eaten. Except for the possibility of allergies, which all corporate and academic researchers and government regulators are attentive to, there are no indications of untoward effects from eating foods from any of the currently harvested genetically modified plants. Nor are there are obvious reasons to worry about the health effects of foods and fibers in the pipeline.

What about the balance between desirable and undesirable effects on the environment, including biodiversity, from insect-resistant GMOs? First, GMO crops require much less chemical insecticide than unmodified crops. According to the U.S. Department of Agriculture, with GMOs the use of noxious polluting chemical pesticides was reduced by one million gallons between 1996 and 1998 (Monsanto, 1999), with concomitant cost saving to farmers. Spraying chemicals indiscriminately eliminates all of the insects in a field, including species that are vital for pollination and biological control. Thus, GMOs can improve insect biodiversity.

However, last year two scientific reports showed that milkweed leaves dusted with heavy concentrations of Bt corn pollen are toxic to monarch butterfly larvae in laboratory experiments (Hansen Jesse and Obrycki, 2000; Losey et al., 1999). This was not surprising, because biologists knew that the Bt toxins were toxic to lepidoptera in general. These findings attracted an enormous amount of public attention and concerns, which were amplified by the well-known fact that there has been an unexplained drop of about 70 percent in the population of monarchs wintering in Mexico since 1996 (Monsanto, 2000).

Is there a relation between the use of Bt corn and the decline in the monarch population? Perhaps. But it's also likely that the effect of Bt corn is relatively small compared with the known effects of habitat destruction in Mexico and the use of chemical insecticides in both Mexico and the United States. More recent experiments, some in the field, indicate that the lethal effect of Bt corn pollen depends on the particular variety of Bt corn—various Bt genes have been introduced—and the level of the toxin the plant produces, as well as the amount of pollen that spreads and how far. A few rows of regular corn between the Bt field and uncultivated surrounding areas can diminish the effects. Wise policy making will have to be based on the factors that effect monarch mortality (chemical insecticide, the spraying of tons of *Bacillus thuringiensis* bacteria, the use of genetically modified pollen) as well as crop yields, costs per acre, and local conditions, such as the abundance of monarchs and the timing of larval feeding compared with pollen production.

Another environmental concern is that pest resistance might be spread through the dispersal of pollen to wild relatives of crop plants, which could lead to insect-resistant weeds. If no wild relatives are in the vicinity, there's no problem. For example, there are no wild relatives of corn in the corn belt of the United States. But in Mexico teosinte grows around cornfields. One technique for minimizing this problem would be to plant a border of unmodified plants around a field of modified plants, the same technique that decreases the exposure of monarchs.

Yet another concern is that insects and other pests might develop resistance to the antipest agent in the GMO. This is already a problem with chemical insecticides and with the alleles that provide spontaneous resistance introduced through traditional plant breeding. The development of resistance to all insecticides is a fact of life for farmers, just as resistance to antibiotics is a medical problem. That's one reason farmers are constantly looking for new ways to control pests. Offsetting measures can and are being taken, such as requiring farmers to plant unmodified crops to inhibit—though not necessarily eliminate—the development of resistance in insects. Since January 2000, the Environmental Protection Agency has required farmers to plant 20 to 50 percent of their acreage in conventional corn (EPA, 2000). Discussions are continuing about whether this is necessary and, if so, what percent of acreage is sufficient, but the general principle is imbedded in the U.S. regulatory structure.

As these examples show, we should not be acting on hunches or preliminary findings or irrational concerns but on thoughtful, informed analysis. In our country, the U.S. Department of Agriculture regulates meat and poultry products, the Food and Drug Administration regulates other foods, and the Environmental Protection Agency regulates pesticides. Approval of crops requires testing for both human and environmental toxicity. The regulatory process must be open, transparent, and vigorously enforced so that the public can judge for itself whether its interests are being served. Evidently this did not happen in the case of the corn flour that was used for making tacos, the story of which was on the front pages of all the newspapers.

To address other issues, science can provide, at best, a modicum of useful information. For many people food is a personal and cultural issue, not a scientific one, and all of us want choices about what we eat. We can also find historical examples demonstrating the occasional folly of some traditional approaches. The French deprived themselves of the nutritious and delicious potato for 200 years after it was brought to Europe from the Andes in the sixteenth century because they believed that potatoes caused leprosy. Tomatoes, another sixteenth century New World contribution to global diets, suffered a similar fate. At first only the Italians were bold enough to challenge the widespread notion that tomatoes were poisonous, as indeed some of its relatives and its foliage are.

Recently a new golden rice has been engineered to produce significant amounts of beta-carotene, the precursor of vitamin A. Scientists hope that, after some additional development, the widespread use of golden rice will reduce the number of people in Asia and Africa who are afflicted with blindness because of a dietary deficiency of vitamin A. Some argue that the golden rice will not be palatable to people accustomed to eating white rice. That's a choice the affected populations must make for themselves. Personally, I find it hard to imagine that people would be willing to watch their children go blind rather than change their eating habits.

The argument that GMOs should not be used because they are not natural is frustrating for scientists. What, after all, is natural in this context? Certainly not our standard diets, which are derived from centuries, even millennia, of careful, directed breeding. Some experts believe that the older breeding methods have achieved about as much as they can in terms of productivity of farmland and water. Thus, in many parts of the world more and more forests are being cleared and more and more land cultivated to feed growing populations. Yet most people agree that preserving forests is essential to preserving biological diversity and limiting global climate change. The new genetic engineering techniques could potentially increase the productivity of agricultural land and water and, perhaps, save forests.

Other aspects of the vocal opposition to GMOs have little to do with science. One motivation for the anti-GMO campaign is antagonism to the practices of large agricultural industries. Some people worry that the commercialization of plant varieties means that they will be unavailable to developing countries, which is a legitimate concern, because about 80 percent of new plants have been developed by companies, though not, I should mention, golden rice. We must try to avoid injustices like those associated with the limited availability of drugs to fight AIDS.

Opposition also comes from the organic food industry, which lobbied hard to include the absence of genetic modifications in the official U.S. definition of organic food, although, in fact, organic farming techniques could benefit greatly from the use of certain GMOs. Other critics of GMOs are people who are honestly concerned about their environmental implications. Some people opposed to GMOs have even become violent. For years, in Europe, they have engaged in the willful destruction of greenhouses, laboratories, and experimental fields; similar acts have occurred in the United States (*Sydney Morning Herald*, June 24, 2000; *Montreal Gazette*, August 10, 1999; *Washington Post*, October 26, 1999; Fumento, 2000).

I do not mean to say that the promoters of GMOs are blameless. Several large corporations have invested heavily, and then promoted, the development and production of seeds of genetically modified plants. The concerned public is naturally suspicious of claims that these plants are harmless and valuable. Suspicions about these claims are reinforced by the fact that the crops have as yet had no direct, obvious advantage to consumers. Can the 6 billion people on Earth (or the 9–12 billion people expected to populate the Earth by 2050) be adequately, economically fed without the investments and products of large companies? American farmers, who are usually pragmatic, initially embraced engineered corn, soy, and cotton because they believe they will be economically advantageous.

Thus far we've seen only the tip of the iceberg of GMOs. Promising research is under way in many areas. Researchers are working on incorporating vaccines—for example, against diarrhea-producing organisms—into edible, easy-to-ship, easy-to-store plants like potatoes and bananas; this could go a long way toward addressing distribution problems for vaccines in many countries. Someday plants may provide fuels and lubricating oils for automobiles, thereby saving fossil fuels and mitigating their damaging environmental effects while making direct use of the energy of the sun. Researchers are also engineering trees to reduce the amount of chemicals needed to produce paper.

On balance, although GMOs can bring real advantages to agriculture, health, and the environment, the use of this new technology has been all but foreclosed, at least for now, in Europe and some other countries. Exaggerated arguments about potential problems—particularly the implication that GMOs are not safe to eat—could bring the United States to a similar position. If scientists can address and allay these concerns, we may all reap a good harvest.

REFERENCES

Ehrlich, P. 2000. The tangled skeins of nature and nurture in human evolution. Chronicle of Higher Education 47(4): B7–B11. Available online at *<http://chronicle.com/free/v47/i04/04b00701.htm>*.

EPA (Environmental Protection Agency). 2000. Biopesticide Fact Sheet: *Bacillus thuringiensis* Cry1Ab Delta-Endotoxin and the Genetic Material Necessary for Its Production (Plasmid Vector pCIB4431) in Corn [Event 176]. EPA 730-F-00-003. Available online at *<http://www.epa.gov/pesticides/biopesticides/factsheets/fs006458t.htm>*.

Fumento, M. 2000. Crop buster. Reason Magazine 31(8): 44. Available online at *<http://www.reason.com/0001/fe.mf.crop.html>*.

Hansen Jesse, L.C., and J.H. Obrycki. 2000. Field deposition of Bt transgenic corn pollen: lethal effects on the monarch butterfly. Oecologia 125(2): 241–248. Available online at *<http://link.springer-ny.com/link/service/journals/00442/contents/tfirst.htm>*.

Losey, J.E., L.S. Rayor, and M.E. Carter. 1999. Transgenic pollen harms monarch larvae. Nature 399 (6733): 214.

Monsanto. 1999. USDA Report Cites Pesticide Reductions and Yield Increases Associated with Biotech Crops. Biotech Knowledge Center, 1653, July 7, 1999. Available online at <http://biotechknowledge.com/showlibsp.php3?uid=1653>.

Monsanto. 2000. Butterflies and Bt corn pollen, Lab Research and Field Realities. Biotech Knowledge Center, 3069, February 15, 2000. Available online at <http://biotechknowledge.com/showlibsp.php3?uid=3069>.

Vorman, J. 1999. USDA issues first estimate of GM crops. Available online at <http://www.gene.ch/genet/1999/Oct/msg00028.html>.

Benefits of Biotechnology

EDWARD A. HILER

Never think for a minute that we are going to build permanent peace in this world on empty stomachs and human misery.

Norman Borlaug
Nobel Laureate

On October 6, 1999, Norman Borlaug, the 85-year-old Nobel laureate, visionary, and father of the Green Revolution, stood in the bright sunlight of a Texas fall day as a new Crop Biotechnology Building at Texas A&M University was named in his honor. In his address to the crowd assembled for the occasion, Dr. Borlaug did not reflect on his life accomplishments. Instead he issued an impassioned challenge for the future. Arguing eloquently that peace will not be achieved until we feed the world, he called on agricultural scientists to pursue advances in biotechnology vigorously to enable humanity to realize this fundamental goal in the twenty-first century (Borlaug, 1999).

As an administrator at a land-grant university, I can tell you that Norman Borlaug's call to advance agricultural science in the service of humanity resonates strongly with our teaching, research, and extension faculty. Yet, on a day-to-day basis, we go into our laboratories and test fields with short-term goals and benefits in mind in the belief that incremental advances will lead us toward solutions to the world's most daunting problems. In this paper, I will outline some of the current and anticipated benefits to humanity from agricultural biotechnology, offer a few examples of exceptionally promising technologies, and note the key challenges we face.

For centuries, farmers have relied on the newest technologies of their era, such as hybridization and selective animal breeding, to produce foods with specific beneficial traits. Biotechnology simply raises that capability to a new level of precision. Like earlier technologies, biotechnology promises to provide many advantages, especially in three broad categories: environmental protection, higher yields, and improvements in human health.

Plants engineered for certain characteristics will have major benefits for the global environment. The primary impact will result from a reduction in the overall use of chemicals to protect against plant viruses, which often claim up to 80 percent of many crops. Like vaccines for humans, biotechnology enables breeders to insert small fragments of plant viruses into crops so they develop natural protection or immunity against the disease and pass this trait on to future generations.

Losses of crops to insect pests can be equally devastating. Biotechnology can confer resistance to those pests in specific crops and locations. For example, crops containing insect-resistant genes from *Bacillus thuringiensis* have made it possible to reduce significantly the amount of pesticide U.S. farmers apply to cotton crops. In the case of cotton alone, the National Research Council reported a reduction of 5 million acre-treatments, or about 1 million kilograms of insecticides, in 1999 compared with 1998 (NRC, 2000). Even though chemicals and their precision application have been greatly improved in the last two decades, residues continue to enter the soil and are washed into watersheds. Biotechnology may offer our best hope for significantly reducing this chemical stress to the environment.

From my perspective as an agricultural engineer whose career has focused on water and soil, the potential impact of biotechnology on the preservation of land resources would be significant. When crops are genetically engineered to resist herbicides, pests, or diseases, farmers can reduce activities that disturb the land. For example, techniques such as weeding require moving the soil, which results in erosion. Engineered crops will make it more likely that producers in both developed and developing countries will retain valuable topsoil rather than sending it by the ton down rivers to the sea.

A second tremendous benefit will be higher yields, which might seem like a disadvantage at a time when U.S. producers are being paid historically low prices for their crops. However, this situation has as much to do with government agricultural and trade policies as with crop yields. Consider the global situation. One of the major technologies that led to the Green Revolution in the 1950s and 1960s was the development of high-yield semidwarf varieties of wheat. Back then, it took plant breeders 10 to 12 years of mixing thousands of genes to produce these varieties. Today, breeders can select a specific genetic trait from any plant and move it into the genetic code of another plant, thereby developing new varieties much more quickly. Speed is not just a convenience for the scientific community; it is critical for us all. World population may reach 9 billion by 2050, and all of those people must be fed. Unless crops produced on land now devoted to agriculture can be made more productive, the disappearance of rain forests, wetlands, and other habitats—and the human misery that come with them—will surely accelerate.

Most crops grown in this country produce less than 50 percent of their genetic potential, and crops raised in the developing world yield far less. The shortfalls are due in large part to the inability of crops to tolerate or adapt to environmental stresses, pests, and disease. For example, in 1993 the disease *Fusarium* caused an estimated $1 billion in damage to wheat and barley crops. Cold, wet weather and frost damage have caused severe damage to the potato crop in the northern plains states. Field crops and the beef industry in the South have suffered from several years of drought. In the developing world, diseases and poor environmental conditions often lead to total crop failure. In response to changes in global climate and the rising demand for food supplies to feed the world, marginal farmland can and must be made productive. Engineering crops to thrive in less-than-optimum soils, with less water, or under other environmental stresses will make that possible.

The third area of potential benefit is improvements to human health. As the connection between diet and health becomes clearer, opportunities will increase to engineer foods that can deliver specific disease-preventing compounds or treat chronic diseases. Extraordinary accomplishments have already been made, such as tomatoes with a higher antioxidant (lycopene) content; transgenic rice with greater production of beta carotene, a precursor to vitamin A; transgenic rice with elevated levels of iron; and fruits and vegetables with higher levels of vitamins C and E. In developing countries, vitamin deficiencies are major causes of blindness and underweight in children, mortality during childbirth, and anemia. Today, these nations must rely on expensive supplementation programs to address nutritional problems. The sustainable solution may well be genetic improvements to food crops.

Another major health benefit is the development of pharmaceuticals and vaccines from transgenic plants. Vaccines for diseases are often expensive to produce, require special storage, and require trained personnel to administer—all major impediments in the Third World. Biologists have produced vaccines in plants, such as potatoes and bananas, against infectious diseases of the gastrointestinal tract. Although these vaccines are in the very early stages of development, they could eventually have a tremendous impact on people in the world's poorest nations. In addition, about one-third of medicines used today are derived from plants. Once genetic modifications are identified, they could increase the yields of these medicinal substances. The frontier of improvements in health may well be in farmers' fields.

A perennial problem for agriculturalists worldwide is accurately diagnosing diseases in plants and animals. Some diseases can be detected visually, but samples must often be sent to a laboratory, and the results may take days or weeks. Delays can be costly when a quick diagnosis could prevent further damage. Two agricultural biotechnologies that will be under development in the years to come are especially important and intriguing—techniques for diagnosing diseases and biochips that can protect against contamination.

Advances in biotechnology are yielding new products and techniques for diagnosing disease. For example, one diagnostic kit under development uses antibodies to detect plant pathogens. This enzyme-linked assay is based on the ability of an antibody to recognize and bind to a specific antigen associated with a plant pathogen. Diagnostic kits have been developed to detect a number of diseases, including bacterial canker of tomato and soybean root rot. These kits do not require laboratory equipment and can be used by producers in the field.

In a related development, scientists at Purdue University and other research laboratories are creating biochips, which mate silicon computer chips and biological assays, to search for proteins, biochemicals, and pathogens in organisms. Researchers could use biochips to discover beneficial compounds in millions of organisms on Earth for use in medicines or to warn of microbial contamination in food supplies.

Other benefits from biotechnology include reduced costs, the jobs and wealth created by a vibrant new industry, and more desirable, fresher, shelf-stable food products for consumers. Biotechnology may not deliver every benefit that we can imagine, but a great number of these and other advantages will very likely materialize. However, these benefits will not occur—or at least not rapidly—unless we address major challenges, including ensuring adequate support for research. *Biotechnology for the 21st Century*, a report from the National Science and Technology Council, highlighted five broad research areas for agricultural biotechnology that merit attention by federal agencies (NSTC, 1995):

- continued mapping and sequencing of animal/plant/microbial genomes to elucidate gene function and regulation and facilitate the discovery of new genes as a prelude to gene modification
- identification of the biochemical and genetic control mechanisms of metabolic pathways in animals, plants, and microbes that could lead to products with novel food, pharmaceutical, and industrial uses
- improvement in our understanding of the biochemical and molecular basis of growth and development, including the structural biology of plants and animals
- elucidation of the molecular basis of interactions of plants and animals with their physical and biological environments as a basis for improving the organisms' health and well-being
- improved food safety methodologies, such as rapid tests for identifying chemical and biological contaminants in food and water

To meet these goals, funding for agricultural research must be on par with funding for health and medical research.

The second challenge facing us is, of course, consumer education. In contrast to the attitudes of European consumers, recent surveys have revealed consistently positive attitudes by U.S. consumers toward biotechnology, if that means producing high-quality foods with health benefits. This confidence reflects the public's trust in the U.S. government's food protection system. That faith might be eroded, however, in light of incidents such as the recent discovery that Taco Bell taco shells contained grain from genetically modified corn that the Food and Drug Administration had not approved for human consumption. Organized interest groups typically use such incidents as fodder for their continued efforts to ban genetically modified foods. Even in some rural areas of Texas, individuals participating in a national campaign run largely through the Internet have brought petitions to county courts asking that the commissioners support a declaration against genetically modified foods. The regulation and labeling of genetically modified foods will clearly be the subject of continued public debate. Given this environment, the agricultural industry, educational institutions, and the scientific community must work diligently to inform citizens about the safe and appropriate uses of biotechnology.

In a land of plenty and in prosperous times, U.S. citizens and their counterparts in Europe have had difficulty understanding why we must pursue advances in biotechnology. I doubt that the billions of people in developing nations have the same concern. It is incumbent upon agricultural scientists and engineers to provide the benefits we know can be realized through biotechnology and to persuade our governments and citizens to support this endeavor. For what is at stake may very well be the future of humanity.

REFERENCES

Borlaug, N. 1999. Speech given at Texas A&M University during the dedication of the Norman E. Borlaug Center for Southern Crop Improvement. October 6, 1999.
NRC (National Research Council). 2000. Genetically Modified Pest-Protected Plants: Science and Regulation. Washington, D.C.: National Academy Press.
NSTC (National Science and Technology Council). 1995. Biotechnology for the 21st Century: New Horizons. Washington, D.C.: U.S. Government Printing Office.

Earth Systems Engineering and Management
The Biotechnology Discourse

BRADEN R. ALLENBY

As a result of the Industrial Revolution and concomitant changes in human population, technological systems, scientific knowledge, culture, and economic systems, we now live in a qualitatively different world from any humanity has previously experienced. The dynamics of many fundamental natural systems—from the grand cycles of nitrogen, carbon, phosphorus, and sulfur to the hydrologic cycle to atmospheric and oceanic systems to the biosphere at all scales—are now dominated by the activity of the human species (Allenby, 1998; McNeill, 2000; Turner et al., 1993). Unless there is a precipitous reduction in the scale of human activity, we must now accept the ethical responsibility for rational engineering and management of human-natural systems.

This will require Earth systems engineering (ESE), which can minimize the risk and scale of unplanned or undesirable perturbations in coupled human-natural systems and, at the same time, manage large, evolving projects and technologies with complex governance, ethical, scientific, cultural, and religious dimensions and uncertainties. Unfortunately, scientific and technical knowledge to support ESE is weak or nonexistent; and the institutional and ethical capacity to complement ESE is, if anything, even more primitive. Therefore, ESE must be considered a capacity that will have to be developed in the coming decades rather than a capacity that can be implemented in the short term.

The technological orientation of ESE reflects the central role of technology as the means by which human cultures interact with the physical, chemical, and biological world. Biotechnology, for example, taken as a general human capability, is a primary means by which we now structure fundamental natural systems. In agriculture, biotechnology has been the most important mechanism by which the anthropogenic world has evolved (Redman, 1999). Agricultural activities throughout history have affected natural systems from the species level to the biome level. The clearing of forests in Europe and North Africa from the eleventh to fourteenth centuries marked the beginning of human contributions to the increase in carbon dioxide, which has affected global atmospheric chemistry. In this sense, ESE is not "new": humans have been engineering Earth systems for centuries. What is new, and has led to the creation of the anthropogenic world, is the scale of human activity and the increasing influence of human activities on natural systems. ESE as applied to biotechnology, and more broadly to the human experience, is, therefore, the assumption by humans of responsibility for what we as a species are already doing. With responsible ESE, we can develop the capability to act more rationally and ethically in the future.

In the context of ESE, "technology" must be understood in its broadest sense as the means by which individuals and human societies improve the quality of life. Technology is the intermediary through which humans affect the physical world and shape their future. The difference between engineering an artifact and engineering an Earth system can indicate the importance of ethics, philosophy, and even theology in ESE. A design team engineering a toaster, for example, works *in an existing cultural and ethical context* that presupposes a market system within which a device to toast bread can be engineered, manufactured, sold, and used, and assumes that society accepts this pattern. The ethical dimensions of the project are explicitly established in legal and regulatory structures—product safety, environmental requirements, and the like. The ignorance of the religious or ethical dimensions of a project is one reason technologists tend to resist the idea that their activities are culturally determined.

The same cannot be said of ESE, which is not an artifact in an existing context; *ESE is the cultural and ethical context itself.* Consider the efforts being made to reengineer the Everglades, a unique biological community, to preserve remaining species and habitat in the face of dramatically increasing human presence in Florida. Designing the Everglades is not just a question of building a dike here or creating a channel there; it entails selecting an objective—for example, continued human presence and some protection for wading birds—that cannot be justified solely on objective grounds. The ethical and, indeed, religious dimensions of the Everglades project are important design objectives and constraints.

Similarly, one cannot think of the engineering of the carbon cycle with the intent of stabilizing climate systems without recognizing that ethical and religious dimensions are critical determinants of the design process. Deciding how to address global climate change—for example, the push by environmentalists to phase out the use of fossil fuels—will have enormous implications for the options available to the human species in the future. The methods selected will necessarily be designed to lead to a certain kind of world—for example, a "natural" world that consumes minimal amounts of materials and energy or a high-technology, rapidly evolving world. For our purposes, it doesn't matter which vision is right;

what is important is that ethical considerations will determine how humans use ESE to influence the evolution of natural systems.

In approaching ESE as a broadly technological discourse, we must be aware of the most crucial difference between science and technology.[1] The goal of science is understanding physical reality and the objective implications of suggested future paths. The goal of technology is to generate options for the future. In a way, it is similar to art. Both are forward-looking activities that not only embody fundamental values but also validate and even create them. As exercises of the human imagination, they not only define present reality but also posit a future vision of the world. Indeed, for the Greeks and throughout the European Middle Ages, art and technology were not differentiated. The Greek word for both was *techne*, meaning art or artifice; even today the British maintain the Royal Society of Arts, Manufactures, and Commerce. Unlike science, therefore, technology (and art) exercises a considerable power that is not widely appreciated (Noble, 1997). Science is concerned with what is; technology creates what will be. The implications are obvious: scientific understanding can be continually tested against reality while technology has far more degrees of freedom.

We can examine some aspects of the anthropogenic world by focusing on biotechnology, one of the most important technologies of our era. Biotechnology raises a fundamental question about the kind of biosphere humans *ought* to design.

The question clearly reflects the ethical and religious dimensions of human experience. Human institutions are not yet ready to address this fundamental question, however. For example, in the time frames being discussed in the climate negotiation process (decades to centuries), the biosphere is essentially plastic at all scales. We already genetically engineer agricultural crops, trees, and bacteria. Even if Europe agrees to forego such technologies, countries like India and China, which must rely on as many technological options as possible to avoid massive civil upheavals and famine, are unlikely to do so. Thus, considering climate change without considering explicitly the potential for innovations in biotechnology would be a victory of ideology over reality and an indication of how far we have to go. In some regions and cultures, powerful groups oppose all genetic engineering and biotechnological activity. Some of this opposition is based on scientific concerns, which can be addressed through additional research. But much of the opposition is ideological. Ideological arguments cannot be resolved through rational discourse, but they are also unlikely to prevail. The power and capability that bioengineering will provide to those societies that adopt it will in all probability ensure that those cultures become increasingly powerful. As a result, the development of biotechnologies is likely to continue.[2]

This leads to another important point. Under traditional international law, only countries are considered competent to make treaties, negotiate agreements, and represent citizens in international forums. Participants in the negotiations on measures to mitigate global climate change, for example, are all nation states; private firms and nongovernmental organizations (NGOs) have been lobbying behind the scenes. However, private firms, NGOs, and communities of different kinds share in the development and implementation of international policy (Mathews, 1997). Formal practice, however, has yet to catch up with this new reality, and the roles of these entities have yet to be defined (Figure 1).

An informal structure of international and regional governance has evolved for several reasons. First, the financial power of many transnational corporations is equal to that of many small countries. In addition, private firms, by and large, are the repositories of technological sophistication. Therefore, they must be primary participants in finding technological solutions to environmental and human rights issues (Netherlands Ministry of Housing, Spatial Planning, and the Environment, 1994). As a corollary, these firms are under significant pressure to include environmental and social dimensions in their performance (Allenby, 1999).

The growing importance of firms is balanced by the growing importance of NGOs. In fact, a number of governments, especially in Europe, rely on NGOs to perform many functions that were formerly performed by governments, such as distributing food aid in African countries. NGOs have also spearheaded many significant environmental and social campaigns, such as the opposition to genetically modified organisms, confrontations over working conditions in factories in the developing world, and sometimes violent attacks on trade and international financial institutions. Polls routinely show that NGOs have more credibility on environmental issues than scientists, private firms, and even government regulators. NGOs have two significant characteristics: (1) they tend to be issue-specific; and (2) few institutional safeguards regulate their establishment. Virtually anyone

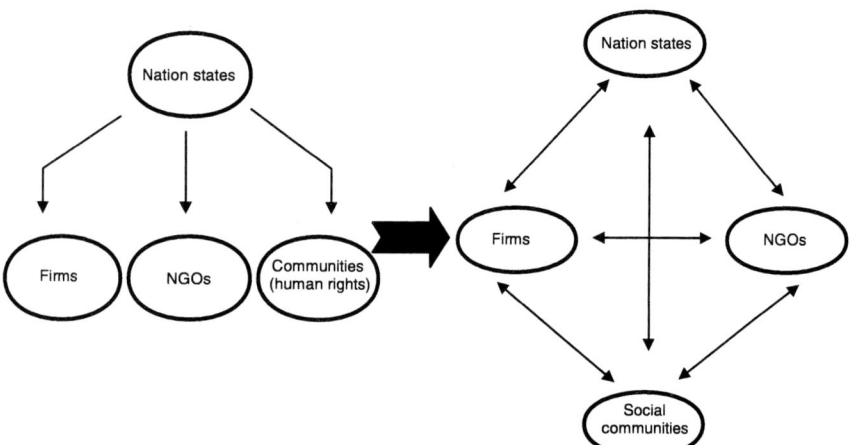

FIGURE 1 The evolution of international governance. Source: Allenby, 1999.

can set up an NGO to represent almost any position within legal constraints. Although this is very democratic, it also means that there are few controls on NGOs that choose to act irresponsibly (Economist, 2000).

The importance of communities—defined by geography or interests—has also increased. In several regions, particularly in Africa, the nation-state structure has not taken hold, and representatives of communities, rather than of the nation state, more accurately reflect the concerns of citizens (Cooper, 1996). Communities affected by a proposed activity, such as the siting of a toxic waste dump, may also participate in governance dialogues if they believe their interests are not being represented. The growth of the Internet and the communications infrastructure has made it much easier for communities of interest to consolidate and represent their concerns in the governance process.

The final element of the anthropogenic Earth is the economic structure. The interesting—and contentious—aspect of biotechnology is that it integrates living systems into the economy. Even a casual overview of the history of agriculture, fisheries, and forestry reveals that this is nothing new. But the scale and emotional and ideological implications are new. With genetic engineering, for example, genes and organisms can be subtly transformed from living things with an inherent value to commodities with a monetary value (in Marxist terms, "commoditized"). This shift has profound ethical and theological implications that are not well understood. However, the example of biotechnology provides a rough framework (Figure 2).

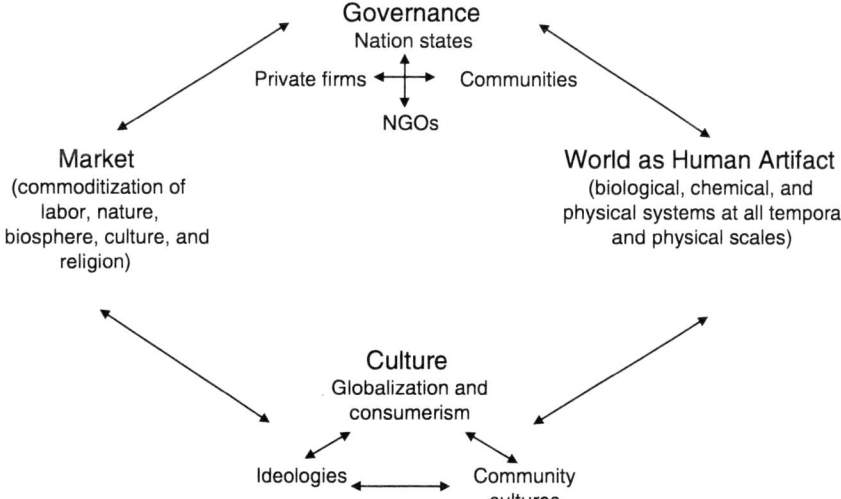

FIGURE 2 The engineered Earth. Source: Allenby, 2002. Used by permission, Darden Graduate School of Business, University of Virginia.

Biotechnology is one of the most important—if not the most important—area of human ESE activity. Biotechnology and the associated economic, political, ethical, and cultural issues provide a lens through which the complexity and challenges of ESE can be clarified.

NOTES

[1] I am indebted to Professor Max Stackhouse of Princeton Theological Seminary for this insight.
[2] This is not to say that all concerns about biotechnology are inappropriate, or that ideological opposition is necessarily wrong in some way. The fact that societies that turned their backs on powerful technologies in the past have been overtaken is not a normative judgment but an objective observation. One need only compare China, where many technologies were discovered but not widely used, and, hence, a technological society never evolved, with Europe, where the Industrial Revolution flourished (Needham, 1991; Noble, 1997).

REFERENCES

Allenby, B. 1998. Earth systems engineering: the role of industrial ecology in an engineered world. Journal of Industrial Ecology 2(3): 73–93.
Allenby, B. 1999. Corporate responsibilities. Nature Biotechnology 17 (Biovision Supplement): BV31–BV32.
Allenby, B. 2002. Observations on the Philosophic Implications of Earth Systems Engineering and Management. Batten Working Paper Series. Darden Graduate School of Business, University of Virginia.
Cooper, R. 1996. The Post-Modern State and the World Order. London: Demos.
Economist. 2000. NGOs: sins of the secular missionaries. Economist 354(8155): 25–27.
Mathews, J.T. 1997. Power shift. Foreign Affairs 76(1): 50–66.
McNeill, J.R. 2000. Something New Under the Sun: An Environmental History of the Twentieth Century World. New York: W.W. Norton & Co.
Needham, J. 1991. Science and Civilization in China. Vol. 2. History of Scientific Thought. Cambridge, U.K.: Cambridge University Press.
Netherlands Ministry of Housing, Spatial Planning, and the Environment. 1994. Netherlands National Environmental Policy Plan Two—The Environment: Today's Touchstone. VROM94059/b/2–94. The Hague: Netherlands Ministry of Housing, Spatial Planning, and the Environment.
Noble, D.F. 1997. The Religion of Technology: The Divinity of Man and the Spirit of Invention. New York: Alfred A. Knopf.
Redman, C.L. 1999. Human Impact on Ancient Environments. Tucson: University of Arizona Press.
Turner, B.L., W.C. Clark, R.W. Kates, J.F. Richards, J.T. Mathews, and W.B. Meyer, eds. 1993. The Earth as Transformed by Human Action: Global and Regional Changes in the Biosphere over the Past 300 Years. Cambridge, U.K.: Cambridge University Press.

Panel III

Engineers and Policy Makers
Partners in the Development and
Implementation of Solutions

Engineers can provide a unique perspective on complex problems. This panel explores the rationale for engineers working closely with policy makers and a growing number of other stakeholders to address global problems.

Gaining a Seat at the Policy Table

ANITA K. JONES

Sustaining Earth systems in the face of projected population growth is a global challenge that will require many different kinds of partnerships. For several reasons, these partnerships will be difficult to forge, difficult to maintain, and will have difficulty being proactive. First, they will involve individuals with very different kinds of expertise, including engineering, ethics, social sciences, legal systems, medicine, physical sciences, politics, and industry. The cultures of these individuals—even their ways of knowing—may differ greatly. An ethicist and an engineer, for example, reason in different ways and from different premises. Therefore, discussions—sometimes extended discussions—will be necessary to ensure effective communication. Second, these partnerships will require cooperation among multinational institutions that typically have a limited understanding, and are even suspicious of, different cultural norms and objectives. These institutions are also subject to outside influences that could force them to change their positions. Third, these partnerships will have to implement plans and investments over decades, and they will have to maintain public support throughout. Long-term international agreements will require not only that nations remain stable but also that they maintain their focus on Earth systems. Governments must be willing and empowered to negotiate and fulfill commitments.

One recent change—the computer—bodes well for communication among people with different kinds of expertise. While science and engineering have long used tools and instruments, computers are now also being used by humanists and other nonscientists. Historians, politicians, and theologians are creating collections of data and, more importantly, using data to validate competing hypotheses and replicate experiments. In other words, they are moving closer to using scientific methods. These shared approaches to problem solving will make it much easier for technical experts and humanists to understand each other.

Unfortunately, I do not see a comparable trend among institutions, which have developed over time in different ways, depending on their cultures. These institutions have very different approaches to problems. For example, one prominent U.S. institution, the Congress, often takes actions that optimize the interests of local constituencies rather than considering global outcomes.

My experience while working at the U.S. Department of Defense (DOD) illustrates these problems. In my position at DOD, I was responsible for the science and technology program. It was well known that the former Soviet Union had invested heavily in military research over several decades and had surely learned some things that the U.S. research community, which had made different investment choices, had not discovered. I, therefore, resolved to improve cooperation between the DOD and the Russian Ministry of Defense—at the level of completely unclassified basic research. Because a substantial portion of military research in any country has civilian applications, I was certain that important areas of completely unclassified research would be easy to find. This had already been done in the area of civilian technologies. The Gore-Chernomyrdin agreement, brokered by Jack Gibbons and others in the White House, had established a U.S.-Russian joint commission on economic and technological cooperation. Why couldn't we also cooperate in military research?

The short version of the story is that, although the United States and Russia did agree to cooperate on a few projects, widespread cooperative research (which also would have moved hard currency into Russia at a time when it was sorely needed) was very difficult to arrange for two reasons. First, because the Soviet Union had always bought and paid for all research and any ensuing development of products for market, no Russian case law had been developed to cover intellectual property—to determine who owns what and when. The Russians had no effective guidelines for evaluating their intellectual property.

For example, DOD offered to pay the Russians to perform experiments with a novel prototype of a cargo aircraft they had built. The aircraft, known as wing-in-ground, can skim the ocean or ground relying on air-ground effect, much like a hydroplaning aircraft. The United States had not built such a prototype because of the high cost, but was willing to support testing of the Russian prototype. At first, following their historical precedents, the Russians felt they could not value the experiments at less than the full cost of developing the prototype. Even after a more reasonable valuation of the proposed research had been made, getting a signed agreement proved to be extraordinarily difficult. The agreement would have represented a partnership between the very ministries that had been in direct conflict for so many years, and even high-ranking individuals were not sure they had the authority to cement such a relationship. In addition, they were concerned that signing an agreement might come back to haunt them.

I suspect that solving our most difficult Earth systems problems will pose similar challenges. Governments with no histories of working cooperatively with other governments, no established processes for conducting negotiations, and no clear lines of delegated authority that allow high-ranking government officials to take action with confidence will have to become partners on long-term projects. So establishing Earth systems partnerships among countries with different legal systems, cultures, economic capacities, and levels of sophistication could prove to be very difficult.

The climate in which we must establish partnerships to address crucial issues of Earth systems engineering is characterized by (1) developing nations that are not yet set in their ways and may be open to new ideas; (2) the prominence of new kinds of organizations; and (3) challenges to the strengths of nation states from global communications and other outside influences.

The bright spot in this scenario is the technical nature of the problems at hand. Because Earth systems engineering is based on technical determinations, technical experts will play a major role in forging international partnerships. Unfortunately, in the United States, engineers and scientists have historically not held many seats at the policy-making table, but we can work to change that situation. In many developing countries, the situation is more fluid, and they could be encouraged to include a much higher proportion of engineers and scientists in high-level policy discussions on Earth systems. A large number of these engineers and scientists have been educated in the United States or other Western or Asian countries and are already our known colleagues.

When opportunities arise, the U.S. engineering and science communities should support more technical participation in policy-making decisions in their countries. When engineers and scientists from developing countries need information and analyses that are available in the United States, the National Academies can support them by offering to share such information to ensure that individuals in high-level positions in policy-making institutions are technically educated. These individuals could advocate policies with a firm basis in engineering and science.

The United States can also build collegial relationships to enable U.S. engineers and scientists to become consultative resources to ensure that engineering advice is based on more information and analysis than a developing country may be able to afford to sponsor. If engineers and scientists can offer sound answers to a wide range of technology-based problems, and not just Earth systems engineering, they could establish their credibility and reinforce support for their participation in policy decisions. A good example of the kind of cooperation that I advocate is described in the 1999 NRC report, *Water for the Future: The West Bank and Gaza Strip, Israel, and Jordan*, which chronicles a border-spanning technical study of sustainable water resources by academies from Israel, Jordan, Palestine, and the United States.

Outreach programs, such as the recent visit by the presidents of the U.S. National Academy of Sciences and the National Academy of Engineering to Iran, is another step in the right direction. The creation of new venues for cooperation helps. Examples include the long standing Council of Academies of Engineering and Technological Sciences and the InterAcademy Council, which is intended to function as an international version of the National Research Council. The National Academy of Engineering also sponsors a series of programs called Frontiers of Engineering, which brings together young engineers in diverse disciplines from different countries to meet and exchange ideas. In addition to meetings in the United States, Frontiers also sponsors bilateral programs with Germany and Japan, and more are on the drawing board.

A variety of other institutions, including nongovernmental organizations and multinational companies, will be participants in Earth systems partnerships. Internet-based virtual communities, which are not limited by national boundaries, may also play a larger role—particularly in the environmental arena. Meanwhile the public, especially in developed countries, is also becoming more active in organized and ad hoc ways. Woe to any corporation that appears to the Western public to be acting irresponsibly.

Because Earth systems partnerships are so difficult to establish and maintain, it would seem wise for us to pay particular attention to problems that can be addressed by limited partnerships, such as issues that are limited to local or regional constituencies. Of course, regional issues can be very complex; witness efforts to reengineer the Everglades or to manage underground water reservoirs in the Middle East. Nevertheless, regional issues may be more tractable than national or international issues. The issue of water seems to be a timely choice to demonstrate the effectiveness of limited partnerships. As difficult as questions about water may be, they involve fewer constituencies, than, say, managing the Earth's carbon cycle, which crosses all national boundaries. Nuclear fallout and airborne pollution don't ask for a visa when crossing national borders. Even as we address these truly global issues, we should give some priority to addressing problems that ought to be slightly more manageable. Half the world's fresh surface water and near-surface groundwater is claimed for some use.

Because water moves slowly, it is a local—or regional—issue. Because it is essential to life, a threat to a nation's water autonomy is an immediate cause for war. The world population is expected to grow from 6 billion to 10 billion in just a few generations, making it critical that water resources be carefully managed. The research community can play a pivotal role in these efforts and can even frame the debate, because the technical community develops the measurement tools, defines the experiments, and wields assessment techniques to determine the status of this Earth system. Analysis of surface and shallow groundwater is relatively well understood, the sensors for determining changes in aboveground and underground reservoirs exist, and the modeling of water ecosystems is reasonably advanced.

The way the research community addresses a problem frames its discussion. Consider, for example, how the subject of carbon dioxide (CO_2) emissions has been addressed in this conference. One presentation was focused on historical data, giving both past and predicted emissions, showing changes in the contributions of generating nations over time and leading one to ask what burden each nation should bear in the future. This inclusive approach raises the issue of remediation not simply in the context of future generations of CO_2, but also in terms of past generations. Discussions that focus solely on future expected emissions imply that all future generators of CO_2 emissions will start from a common position, which is obviously not the case.

In addition to framing the debate, the engineering community maintains "engineering ethics." Young engineers are trained to make choices to ensure safety and integrity within the constraints of the project at hand. This is an example of microethics. Macroethics would entail determining the ethical implications of an entire engineering project on the Earth as a whole. To ensure that engineers are trained in macro- as well as micro-ethics, our educational programs must teach skills for solving problems, not just at the technical level but also through international cooperation. Engineering design courses should deal not just with the design of a widget but also with the policy implications of interactions between engineered systems and the global environment. (For a more detailed discussion of micro- and macroethics, see [*The Bridge* 31(4):35–38], Bill Wulf's address at this year's Annual Meeting of the National Academy of Engineering.)

Finally, engineers can perform the research and refine the technologies that can contribute to Earth systems solutions. We can build the necessary tools and techniques, particularly assessment techniques, including developing models of Earth systems and setting standards for collecting and assessing data.

The United States has a fine record of stepping up to meet new challenges by mobilizing government funding and the resources and talents of the technical community. When the nation was young, states such as Virginia and Illinois invested in canals, and later regional railroads, to stimulate commerce. And President Lincoln and Congress allocated bonds and land grants for the construction of the railroad connecting the East to the Pacific. That funding combined with innovative and heroic engineering created America's transcontinental railroad. And during World War II, the science and engineering community provided new knowledge and technology that materially contributed to winning the war, again with government funding.

Earth systems problems pose a global, rather than a national, challenge. This nation must once again mobilize sustained funding and long-term technical activity to address these problems. Just as we have risen to meet past challenges, we must formulate an approach for meeting present challenges. But engineers must be sitting at the policy-making table to have the greatest effect. Too often engineers and scientists have been up in the gallery looking down when policy decisions were made, rather than helping to frame the policy debate. We also must help our colleagues in developing nations to secure seats at their policy tables.

Successful Public-Private Research Partnerships

KATHLEEN C. TAYLOR

Industry must assume lead responsibility for the development, commercialization, and global dissemination of the technologies needed to meet the environmental, energy, and economic challenges of the new millennium. However, because there is often little or no market pull for more costly technologies that address broad-based environmental concerns, such as climate change, governmental policies, initiatives, and research can and must play a major supporting role.

Innovative public-private partnerships have emerged as one important form of government support for accelerating the development of new technologies. These public-private collaborations can significantly expand the breadth and depth of technical expertise available to the individual partners, reduce the costs and risks of research and development, and bring new technologies to the marketplace faster. The federal government can also provide incentives to promote new technologies and support policy objectives through tax policies and cosponsored research.

Take the Partnership for a New Generation of Vehicles (PNGV),[1] for example, an innovative, successful public-private research partnership begun in 1993 to further some extraordinary policy objectives. The program objectives include reducing imports of foreign oil and restoring our balance of trade by improving the energy efficiency of U.S. vehicles. The partnership, therefore, has three interdependent research goals.

The first is to improve significantly the national competitiveness of U.S. auto manufacturers. The second is to implement commercially viable innovations from research on conventional automotive vehicles. The third is to develop a revolutionary new class of vehicles that can achieve fuel economies of up to three times those of 1994 family sedans while maintaining comparable performance, size, utility, and cost of ownership, and meeting or exceeding federal safety and emissions requirements. The third goal is not achievable with existing internal combustion engines; the intent was to force the development of radical changes in vehicle materials and power systems.

Under PNGV, the government and private engineering communities work together, supported by the U.S. Department of Energy, to develop technologies and achieve the overall objectives. The collaboration of the auto companies and the federal government in a nonadversarial environment is the most critical factor in the success of PNGV.

Several other elements of this partnership have also been important to its success. First, the goals of the program are significant and compelling, and, therefore, they attracted initial resources and technical talent to the program. In addition, top management actively participates in the program. Second, PNGV has been conducted with full awareness of market forces and the diverse resource base required to move beyond traditional automotive technologies. Third, the relationships among the partners are clearly defined, and an effective organizational structure facilitates program management; intellectual property rights were established at the start. Finally, the program has maintained a high level of accountability, through concrete technical milestones and deliverables for measuring progress and annual external monitoring by a panel assembled by the National Research Council.

As PNGV shows, successful public-private partnerships must maintain transparency and accountability to avoid allegations of "corporate welfare." They also require managerial and budgetary flexibility to adapt to changing technical and economic conditions throughout the life of the program.

Further innovations in public-private partnerships would enable them to address a range of policy-driven technical objectives:

- the development of new technologies to support stringent regulatory requirements
- meeting commercialization requirements for new technologies that have limited market pull but that would provide substantial public benefits
- the development of new technologies that require large capital investments in new facilities, communications, transportation, or other infrastructures
- global diffusion of existing new technologies with substantial environmental benefits, such as cleaner, more energy-efficient infrastructures for developing countries

NOTE

[1] Information on PNGV is available online at <http://www.ott.doe.gov/oaat/pngv.html>.

Defining What We Need to Know

DANIEL R. SAREWITZ

How do we know what we need to know? This is not a question that we as a society approach in a particularly rigorous manner. When a technical problem rears its head in a political way—anything from global climate change to the AIDS epidemic to nuclear waste disposal—we are often quick to call for more research but not to define the problem we are trying to solve. Our casualness about defining research problems has often meant that we conduct world-class science that is ill-suited to addressing the problem at hand. Sometimes our approach to research can complicate the political challenges surrounding a problem.

New knowledge is usually a good thing, but all new knowledge is not equally useful; in some cases, existing knowledge may be sufficient to enable action. In the areas of global climate change and human health, for example, research agendas have strongly focused on advancing our fundamental understanding of climate dynamics and molecular genetics. But we haven't really defined the particular problems we are trying to solve or the types of knowledge best suited to enabling progress. I believe one mistake we often make is to define such problems as scientific in nature, when in fact it would be more productive to consider them, at least partly, as engineering challenges. This mistake reflects a widely held predisposition that fundamental understanding must always precede action. This assumption may reflect the political strength of the basic research community in the science-policy pecking order, but it does not do justice to the complexities of the real world. To illustrate my point, I want to reframe two well known and well understood challenges concerning the environment and the developing world. Let me begin with a story.

The term Lupang Pangako means promised land—the sardonic name given to a garbage dump outside the city of Manila inhabited by almost 100,000 people. I visited Lupang Pangako about 15 years ago in a different life as a geologist, and the place really is hell on earth. As you drive through the Promised Land, you see stygian mists rising from the hillsides, the mountains of garbage, and if you look closely you see movement everywhere in the distance. You soon realize that the mountains are covered with people scavenging for their livelihoods.

You may remember that in July 2000 torrential typhoon rains caused a huge landslide in the Promised Land that buried more than 200 people under a mountain of garbage. To me, this horrific event provides a powerful indicator of how we should be thinking about the impacts of climate on people and about human adaptation. The problem was *not* whether the typhoon was an above-average or below-average event. It was not a problem whose root causes could be revealed through a better understanding of anthropogenic climate change. The problem was that 100,000 people were living in poverty so deep that they could survive only by culling garbage.

The results of humanity's mistreatment of the environment fall disproportionately on poor people, on developing countries, and on tropical regions. Although these impacts are most severe in their chronic forms, they are most spectacular in their catastrophic versions, such as this landslide. As Figure 1 shows, the number of disasters has risen sharply throughout the world in the last 30 years, most markedly in the developing world. This trend does not reflect a changing climate; it reflects changing demographics—growing numbers of poor people living in urban areas, living in coastal regions, living on garbage dumps. Unlike changes in climate, this trend is something we can control. These are not natural disasters; these are intersections of natural phenomena and complex sociopolitical and socioeconomic processes.

The number of disasters will continue to rise because we know that demographic trends are pointing toward more urbanization and greater numbers of impoverished people moving from agrarian areas to cities—often to areas in harm's way. Megacities like Jakarta and Manila that have nearly 10 million people apiece are subject to typhoons, volcanoes, earthquakes, landslides, epidemics, and floods, for example. Because generating more knowledge on climate dynamics cannot help us in the short term, it is worth talking not just about the behavior of the climate and our capacity to modify it by reducing greenhouse gas emissions, but also about the interactions of social systems with climate and the engineered systems that sustain human beings. These systems are not sensitive to emissions of carbon dioxide but are very sensitive to demographic and socioeconomic trends. We have much less control over the future behavior of the climate than we do over the behavior of human beings.

Given the complexity of these interdependent systems, the practical challenge is to learn to operate in ways that minimize our impact on the planet and maximize our resilience in the face of unpredictable events and the ever-changing

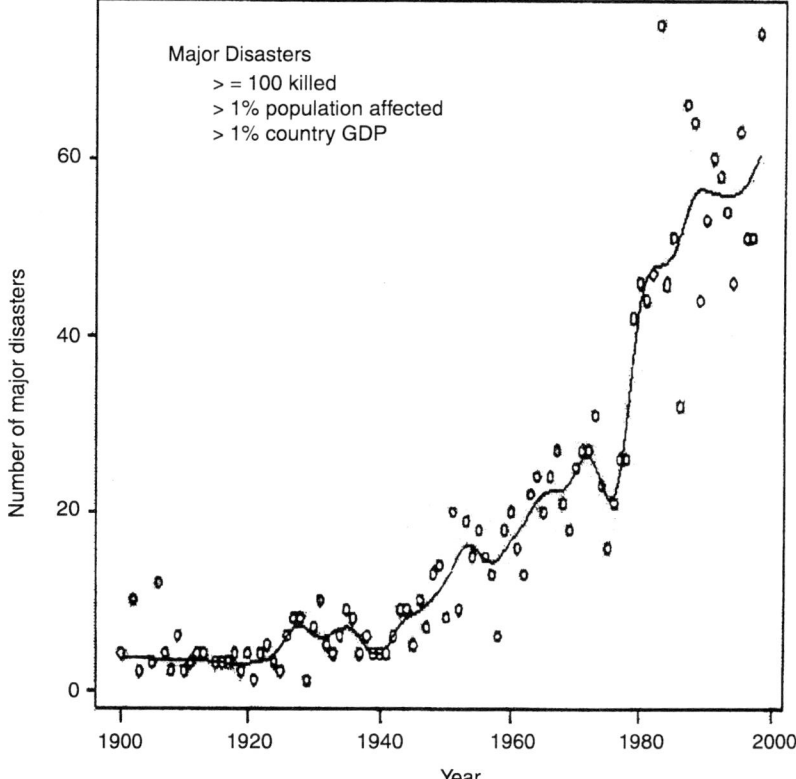

FIGURE 1 The rising number of disasters. Source: Office of Foreign Disaster Assistance/The Centre for Research on the Epidemiology of Disasters, 2000.

relationship between humanity and the environment. But the key is to focus on the human condition, because, in a world of 6 billion or more people, that is the central variable in the relationship between society and environment. This focus will require both much less and much more of engineering than we have asked in the past: much less because the idea of complete knowledge of, and control over, nature has been revealed as illusory; much more because we also know that, in the absence of complete knowledge and control, our most feasible course is to learn by doing—and doing means engineering.

Until recently, efforts to connect climate research to specific social outcomes, such as reducing people's vulnerability to climate and weather and improving management of water resources, have been afterthoughts at best, modestly funded—typically at a level of about 5 percent of total program support—and poorly integrated into larger programs. However, researchers on the human dimensions of climate change are beginning to understand how to generate knowledge that improves the capacity of people and organizations to respond to a dynamic climate. Much of this knowledge will have to come from Earth systems engineering (ESE).

The social outcomes of science and engineering do not emerge fully formed from the laboratory; they are created by evolving interactions between the results of research and the needs and capacities of society. One central concept of ESE is that many complex problems must be framed in an integrated technical and social context. This is both a technical and a political insight. Just as no single discipline can capture the complexities of the interactions between natural and social systems, neither can any single perspective provide a vision that is responsive and accountable to diverse stakeholders. The ESE approach is to develop a repertoire of tools that can be applied as needed to move toward a vision. Both the approach and the vision will change over time. A solution to a problem, therefore, is the formulation and implementation of an integrative and iterative *process*, using a combination of social and technical approaches. A solution is not a static *condition* of complete control—such a condition is impossible.

Now I want to turn to the public health effects of poverty and environmental degradation. This subject may seem to be far from an environmental engineering issue, but I maintain that it can productively be considered in that light, and that the failure to do so reflects our inability to "know what we need to know." Figure 2 shows the relationship between GDP per capita and life expectancy. On

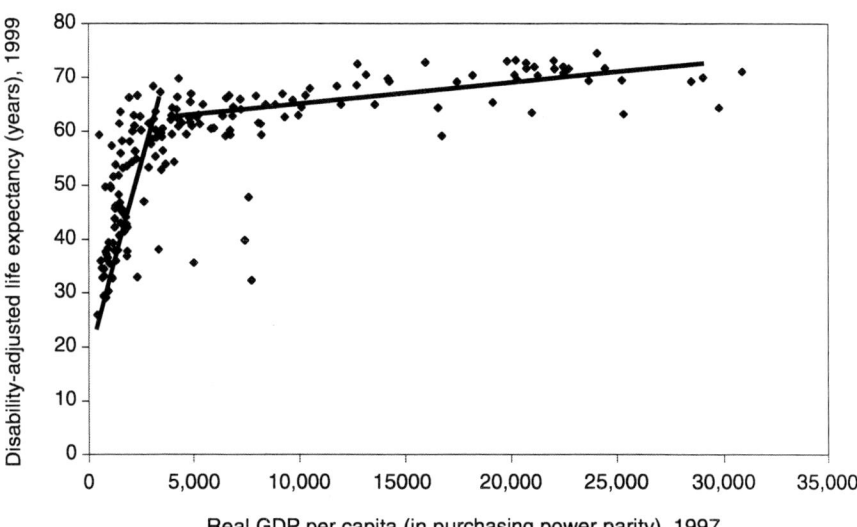

FIGURE 2 GDP per capita versus disability-adjusted life expectancy. Source: World Health Organization, 2000.

the left side is disability-adjusted life expectancy; these are the years one can expect to live without a serious disability or hospitalization.

About 80 percent of the world population resides on the very steep part of the curve, where there is a strong correlation between increasing GDP per capita and increasing life expectancy. The direction of causation runs both ways—more wealth may enable a person to have better health, but better health also enables a person to be economically productive. Yet the majority of biomedical research dollars are aimed at diseases that mostly afflict people on the flat part of the curve, where the correlation is very weak. GDP per capita rises to great heights, but life expectancy remains fairly level once one's income tops about $5,000 per year. Even if a health problem afflicts people on both ends of the curve—AIDS, for example—the interventions useful for people living in the flats may be irrelevant to those living on the slope, as we have discovered to our shame in Africa.

The problems faced by nations on the steep part of the curve cannot be best addressed through biomedical research; they are often problems of Earth systems engineering. They are about finding innovative, efficient, and affordable ways to deliver water, sanitation, food, basic health care, safe shelter and workplaces, and higher environmental quality—that is, making it unnecessary for people to live on garbage dumps. These kinds of challenges require putting some basic infrastructure elements into place. Although this is fundamental, on-the-ground engineering, past failures attest to the extreme difficulty of implementing changes. We must ensure that the technical criteria for engineered systems are compatible with the social, political, economic, and environmental contexts in which they must operate. This is an enormous challenge for ESE.

As these examples of the interactions between public health and climate change illustrate, science policy is often not well aligned with social needs—even if proponents have explicitly invoked those needs to justify the science in the first place. This misalignment reflects a failure to think carefully about what we need to know to address a given problem, a failure to define problems in a way that can stimulate positive action. Some problems we have defined as scientific might be reconsidered as engineering challenges. The obvious justification for this reevaluation is that we already know enough to take action *now* on some problems, such as climate impacts and public health. But something more subtle is also going on. By redefining a difficult environmental challenge as an engineering problem rather than a scientific problem, we acknowledge we must inevitably take action in the face of uncertainties and complexities. This is what engineering is all about—finding solutions that can work despite our imperfect knowledge. This is also, interestingly enough, what politics is all about.

The central point is simply this: setting effective research priorities for solving environmental problems requires that we think carefully about what we really need to know to take action. This will require an explicit articulation of the social goals we are trying to achieve. In the absence of such clarity, we often define environmental problems in terms of scientific opportunity rather than real-world issues. Not surprisingly, this can lead to distortions—spectacular climate-modeling capabilities paralleled by growing global climate change, for example. In this context, it seems to me that a new and clearer look at our most serious environmental problems demands a more central role for Earth systems engineering.

REFERENCES

Office of Foreign Disaster Assistance/The Centre for Research on the Epidemiology of Disasters. 2000. EM-DAT: The OFDA/CRED International Disaster Database. Available online at <http://www.cred.be/emdat/intro.html> (December 6, 2001).

World Health Organization. 2000. World Health 2000. Geneva: World Health Organization. Available online at *<http://www.who.int/whr/2000/en/report.htm>* (December 6, 2001).

Panel IV

Designing the Urban Centers of Tomorrow

In the next several decades, cities—major hubs of energy, resources, and technologies—will experience tremendous growth. This panel identifies some critical challenges facing today's cities and discusses potential innovations for designing the urban centers of tomorrow.

Rethinking Urbanization

GEORGE BUGLIARELLO

Since the emergence of the first concentrated human habitats some 10,000 years ago, urbanization has increased vertiginously. Even if not everyone agrees on what exactly constitutes a city or an urban area, most people agree that rapidly increasing urbanization is a new and seemingly uncontrollable phenomenon. At the beginning of the twentieth century, only about 5 percent of the world population lived in urban areas. Today, the figure is 40 percent and is projected to increase to 60 percent in the next 20 years. In the United States, the percentage will be even higher. If current trends continue, by 2030 all of the world's population growth will be in urban areas. Over the next 30 years, urban population will increase from 2.9 billion to 4.9 billion people, concentrated mostly in developing nations. The greatest population growth will occur in Asia, but Africa will have a higher rate of growth. The number of cities with 5 million inhabitants will increase from 41 to 59, and the number of cities with 10 million inhabitants (called megacities) will increase from 19 to 23, mostly in the developing world (Brennan-Galvin, 2000).

Urbanization is the most powerful and most visible anthropogenic force on Earth, affecting the surface of the Earth, the atmosphere, and the seas. The expanding surface area occupied by cities and the resources required to supply their needs are absorbing or transforming, directly or indirectly, increasing amounts of forests and arable land. Because cities are virtually devoid of oxygen-generating vegetation, they exacerbate the problems of atmospheric pollution. The surface "footprint" of a typical city consists predominately of buildings and concrete or asphalt, all of which repel water and can lead to deprivation and even subsidence of aquifers. Aquifers under Mexico City, for example, have dropped some nine meters since the beginning of the last century (Rowland, 2000). In the developed world, the extensions may be hundreds of times larger than the surface area of the city they supply, and material and energy resources are consumed at rates per inhabitant an order of magnitude greater than those of cities in the developing world.

Simple urban agglomerates emerged about 10,000 years ago, made possible by the invention of agriculture. Substantial cities began to emerge about 5,000 years ago and, on a greater scale, later, with cities like Memphis, Babylon, Athens, Beijing, and Rome. In the vast period of time between the development of agriculture and the Industrial Revolution, most innovations, such as codified laws, organized armies, and bureaucracies, occurred primarily in the social domain, although some crucial new technologies, such as aqueducts, bridges, and fortifications, also emerged.

Since the Industrial Revolution, waves of technological invention and innovation have succeeded each other with increasing rapidity making cities what they are today. Industrial manufacturing attracted armies of workers to cities; railroads, and later airports, weakened the commercial advantages of maritime cities; the internal combustion engine helped create suburbs; electricity made all sorts of labor-saving devices possible; the elevator made vertical expansion possible; sanitation made cities healthier; radio, later complemented by television, computers, and the Internet, enabled people to interact without physical contact and to work cooperatively at a distance (Moss, 1998). The emergence of biotechnology and biomachines will affect future cities in ways we cannot fully fathom.

The pace of change increased dramatically over this same period of time. More than 100 years separated the Industrial Revolution from the invention of the internal combustion engine; only 50 years separated the radio from the computer; and about 30 years separated the computer from the emergence of biotechnology. These innovations have added to the fascination and the promise of cities, whether realistic or not, and have fueled the continuing growth of most urban areas. No matter how undesirable and ultimately unsustainable cities may be, thanks to technology and the seeming availability of land and other critical resources, there seems to be virtually no limit to the growth of cities (Groat, 2000).

THE CITY ESSENTIAL BUT DYSFUNCTIONAL

Today's cities are essential sources of opportunities for social advancement, the creation of wealth, globalization, and creativity, are outlets for psychic energy, and are also instruments of reducing birth rates. But many cities are also dysfunctional both environmentally and socially. They are large consumers of resources, reservoirs of poverty, and concentrated sources of pollution. Cities are also congested and, especially in the rapidly growing megacities of the developing world, bursting at the seams. They are difficult to manage, particularly if a scarcity of natural and financial resources compounds problems. And they pose risks to their inhabitants.

Cities affect their environments by drawing resources—materials, air, water, and energy—from increasingly long distances (the resource "footprint"). Their products tend to be distributed worldwide and become sources of pollution elsewhere. City-genic pollution on the ground may be limited to a few hundred miles, but air pollution may circle the globe. Cities affect their environments regionally because they cover more and more surface area, make intense use of their hinterlands, and encroach on coastlines.

Pollution in large urban aggregates is aggravated by traffic from commuters who must negotiate sometimes long distances between their residences and their places of work and by the increasing use of heating and air conditioning. The concentrated nature of cities reduces the space available between dwellings, as well as the space inside dwellings, precluding less-polluting remedies, such as higher ceilings and shading by trees, which were (and are) available in less-dense habitats. The increase in poverty, particularly in cities of the developing world, is a disturbing trend associated with urbanization. Poverty adds to the dysfunction of cities and often contributes to urban sprawl by encouraging the flight of more affluent inhabitants from the city core.

Risks associated with urbanization are attributable to natural hazards and anthropogenic causes, or a combination of the two. Natural hazards, from earthquakes and floods to volcanoes and diseases like malaria, have been increased by the heedless expansion of cities in threatened areas. Anthropogenic risks include accidents, war, terrorism, crime, changes in the economy, and lifestyle diseases, such as depression, bronchitis and emphysema, tuberculosis, and AIDS. Cities today are vulnerable to unprecedented risks of terrorism and weapons of mass destruction (Figure 1).

Slums and congestion, such as overcrowded roadways and air traffic delays at major airports (estimated to have cost the United States some $5 billion annually more than a decade ago [Craig, 1988]), are ubiquitous signs of urban dysfunction in both developed and developing countries. Another sign of dysfunction is the difficulty of disposing of solid waste, a problem that could be addressed in many creative ways but generally remains, particularly in the developing world, one of the most intractable problems of cities. More subtle signs of dysfunction are urban sprawl, the monotonous grid pattern of streets, and monocultural zones devoted exclusively to a single kind of activity, such as shopping malls or financial districts. These areas are usually deserted when those activities end.

A problem common to all but the most affluent cities is that many elements of their infrastructures have not been extended or improved since they were built. Because of rapid expansion and high cost, upkeep on railroads, bridges, sewers, water mains, major roads, and buildings have not kept pace with urban growth.

MANNHEIM 1695

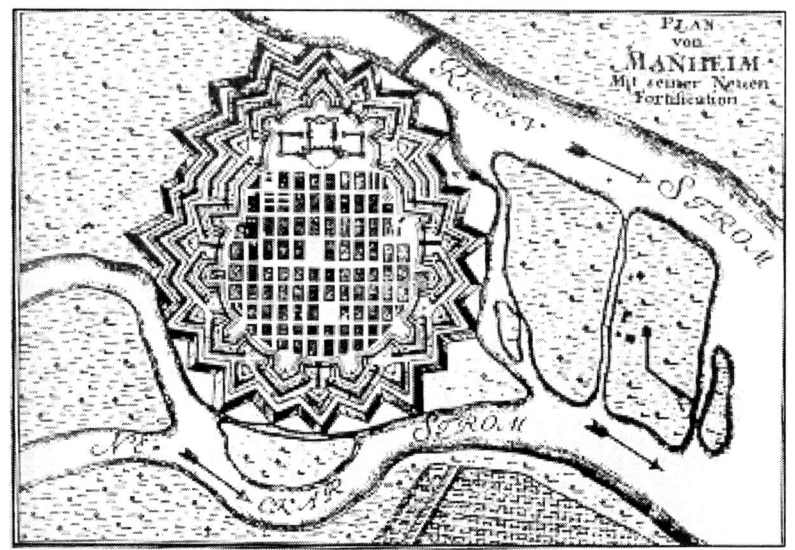

a

WORLD TRADE CENTER 2002

b

FIGURE 1 a. The city protective (Mannheim 1695). b. Ground Zero (New York, 2001). Sources: Rykwert, 2000; Huntley, 2001.

THE CHALLENGE

For cities to be environmentally and socially sustainable, we must rethink our ideas about them. For a long time, the design of the city of the future has been the focus of passionate debate, focused on utopias, ideologies, theories, and experiments. In fact, this debate has been going on for centuries. Leonardo da Vinci's plan for separating pedestrian and vehicular traffic (Figure 2) and the garden city proposed a century ago (Perry, 1929; Relph, 1987) are two ideas that are still relevant; many other concepts have not stood the test of time. We are now at a moment of great urgency and great opportunity. On the one hand, we are faced with the urgent problems of explosive urban growth. On the other hand, new technologies are now available that offer hope for the future of cities, from information technology to new capacities for designing and constructing infrastructure. The time is right for us to use these new technologies and to press for the development of new ones to help us realize our goal of enhancing the positive attributes and reducing the dysfunctional attributes of our cities.

LEONARDO'S DESIGN

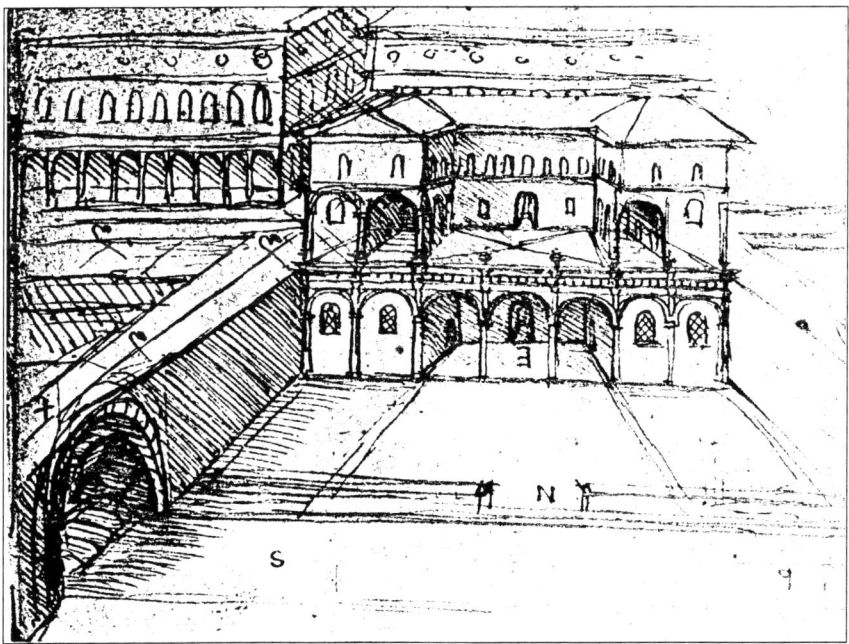

FIGURE 2 Leonardo's sixteenth-century design for separating pedestrian and vehicular traffic. Source: Istituto e Museo di Storia della Scienza, 2001.

PRAGMATIC IMPERATIVES

Cities are extremely complex organisms, and their future forms cannot be projected in detail or prescribed in advance. We can, however, identify some essential characteristics of viable future cities. Regardless of ideology, most would agree that cities of the future, in America and elsewhere, must respond to certain pragmatic imperatives: hazards to inhabitants must be reduced, livability improved, and sustainability ensured. Also, cities must be capable of existing indefinitely in time without causing irreparable damage to the environment.

The city of tomorrow must be a caring and emotionally satisfying place to live; it must be ecologically sound; it must make intelligent use of its resources and technology; and it must be manageable (Figure 3). These characteristics should interact synergistically in response to the imperatives (Figure 4). Thus, to improve livability, a city should be caring and emotionally satisfying, which, in turn, implies that a city must be intelligent, manageable, and ecologically sound. To be sustainable, a city must be ecologically sound. To reduce hazards to its inhabitants, it must be intelligent and manageable. The elimination of slums will require synergy between the "city ecological," the "city manageable," and the "city caring and emotionally satisfying." Similarly, reductions in consumption will require synergy between the "city efficient" and the "city manageable." These synergies will not be easy to achieve, but the problems of today's dysfunctional cities cannot be remedied without them.

The City Caring and Emotionally Satisfying

The city caring and emotionally satisfying will provide jobs, housing, health care, and educational opportunities, give its citizens a sense of protection, and recognize poverty as an urgent problem. Poverty threatens a city's physical and emotional health, and many consider eliminating it essential to sustainability (Perlman, 2000). But eliminating poverty will not be enough. A sense of belonging, a sense of pride, and a sense of adventure are also essential ingredients of the city caring and emotionally satisfying. Such a city must also ensure stability (changing the current policy of constant tearing down or reconstructing, making cities architectural palimpsets), be aesthetically pleasing, and be well managed— in other words, it must be not only functional but also beautiful. To satisfy a sense of adventure, the grid layouts we inherited from the ancient Greeks and Romans and the extreme segregation of functions in separate quarters of a city must be avoided. Today, for example, impersonal gleaming towers of the business district often leave no room for diverse, small-scale activities.

The City Ecological and Sustainable

A future city that does not cause irreparable ecological damage and is sustainable must limit, or even reduce, its geographical and resource footprints. The

The City Caring and Emotionally Satisfying
- Community participation
- Adequate jobs, housing, and health
- Sense of belonging
- Sense of pride
- Sense of adventure
  . . . .

The City Ecological (Sustainable)
- Contained geographical footprint
- Reduced resources footprint

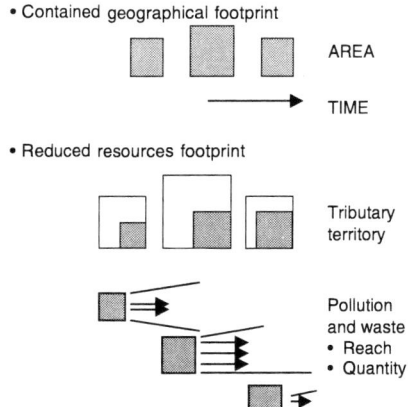

The City Intelligent
- Self-adapting
  - Speed of response
  - Adequacy of response
- Efficient
  - Traffic control
  - Flexible scheduling
  - Reduced waiting time
  - Resource efficient
  - Reduction of poverty
  - Education

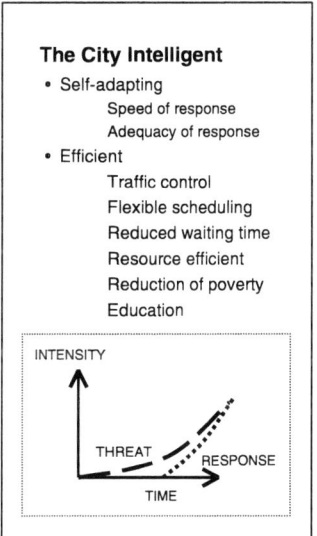

The City Manageable
- System of systems (neighborhood-cluster-city)
- Localized vs centralized activities
- New organizations and services (public/private)
- Control of technology
- Community participation

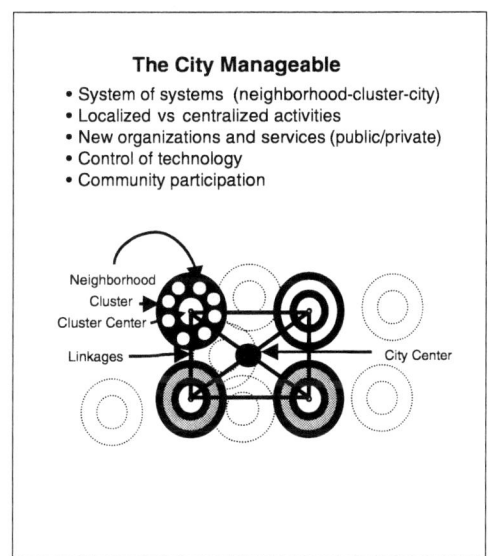

FIGURE 3 The city of the future: caring and emotionally satisfying; ecologically sustainable; intelligent; and manageable.

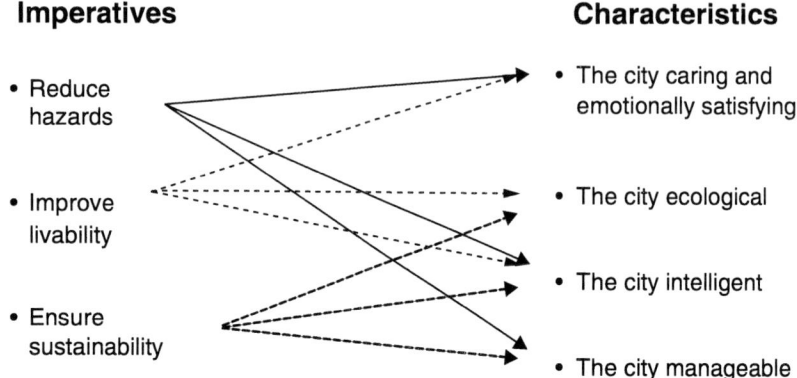

FIGURE 4 Pragmatic imperatives and the necessary synergies.

area occupied by a city and the tributary territory necessary to feed and otherwise support it cannot continue to grow proportionally to the city's population or affluence. Reducing the resource footprint will also mean reducing the size and intensity of the plume of pollution and waste emanating from the city. Because the city is an accumulator of substances, recycling and "mining" those substances must become important sources of materials for the city of the future, thereby reducing the city's resource footprint (Graedel, 1999). The city ecological will rely, as much as possible, on natural means, both biological and energetic (Lewis, 1998). Wetlands can be used to reduce wastewater treatment, and renewable, conservative energy sources, such as wind and solar radiation, can be used to reduce energy demands. Today's conservative energy sources may not be sufficient to satisfy the needs of cities, however. We must be aware of the risks of their overdevelopment, which could create some new ecological stresses.

The City Intelligent

A city intelligent must have the ability to adapt to change. The new capabilities that enhance a city's ability to adapt will include sensors, geographic information systems, improved telecommunications, the ability to simulate and assess trends, and a nimble management structure. A city intelligent must also be efficient in its use of all kinds of resources, including human resources. For example, it could take advantage of advanced traffic control systems and flexible scheduling of city activities to reduce congestion. Education is essential to the city intelligent and efficient, not only traditional education, but also an education in living appropriately in the city—learning how to behave in crowded situations and in traffic, how to reduce pollution through changes in behavior, and how to participate effectively in community decisions.

The City Manageable

A manageable city will balance localized and centralized city activities. No matter how large or small the population of a city, information and telecommunications technologies have become essential to making a city manageable and to enhancing community participation. The city manageable will also endeavor to guide the use of its technologies and encourage the creation of technologies that respond to its needs, rather than merely reacting to new technologies. A good example is the misuse of the automobile, which has a very large footprint, demands that a large portion of the city be devoted to parking, and causes congestion. The city manageable will stimulate the development of new technologies to address the problems caused by automobiles, including environmental problems and parking.

Regardless of its physical configuration, the city manageable must be governed as an organic whole that cannot be rigidly planned but can be guided in desirable directions. Approaching a city as a complex "system of systems" would be a useful guide for organizing services, transportation, utilities, and other aspects of the urban environment and identifying relationships among individual neighborhoods, larger neighborhood clusters, and the city as a whole (Gallopin et al., 2001).

Considering neighborhoods as the organizing units of the city is not a new idea, but it continues to make sense for the city of the future. Walkable neighborhoods, for instance, reduce congestion by encouraging the creation of a hierarchy of transportation hubs connecting the city's components. With wise public planning, suburban sprawl can be stayed in American cities, discouraged in the more concentrated European cities, and avoided in many exploding cities of the developing world. In principle, it should be easier to devise entirely new organizations and systems for cities in the developing world, many of which are expected to double in population in the next 15 to 20 years. Unfortunately, these cities often lack the resources to make those changes. It will be even more difficult to make radical changes in the mature cities of the developed world, however, even though they have the necessary resources.

Other challenges for the manageable city are the role of self-help and sweat equity in housing poorer segments of the population, the development of financial instruments, such as public-private partnerships, to encourage entrepreneurship, home ownership, and economic development, and the pooling of resources and markets with other cities. The relationship between city policies and national policies—including policies that encourage alternatives to concentrating growth in larger cities—is an important challenge.

Another aspect of the challenge of making a city manageable is dealing with unrealistic expectations in an era of burgeoning technological possibilities. Overly high expectations can affect the stability of a city and may even have a global impact. The problem is particularly acute in the developing world. The city manageable must think in terms of reaching rapidly growing areas with essential services by devising "good enough" solutions as opposed to constructing the kind of costly traditional infrastructures that were developed in affluent, older cities.

THE BIOSOMIC CITY

Fundamentally, a city is a complex bio-socio-machine that I call a "biosoma" for short (Bugliarello, 1998, 2000). The biosoma is an entity created by the interaction of: a biological component (human inhabitants, as well as other forms of life, such as vegetation and microorganisms); a social component (the ensemble of collective activities, ideas, and organizations of the inhabitants); and a machine component (the tangible and intangible artifacts that support the life of the city). Each component of the biosoma has distinctive influences on the function and design of the city. The biological component can self-replicate and can be recycled by nature (e.g., through microbiological processes); these capabilities are essential to the sustainability of the city. In addition, human emotions and feelings play a crucial role in the creation of a city caring and emotionally satisfying. The characteristics of the machine component include reliability, precision, and power, but also inflexibility. The characteristics of the social component fall somewhere in between. Like the machine component, the social component increases the reach of the individual and may also embody reliability, precision, and power (e.g., in social organizations, such as bureaucracies), but it also harbors collective feelings and emotions that can erupt with unforeseeable consequences.

Striking a balance among the three biosomic components is important to maintaining a city's positive characteristics and reducing its dysfunctions. The key to an effective balance is to ensure that the human biological component is not overwhelmed and made to feel powerless by the infrastructure or by the social organization of the city. For example, there must be a balance between bioremediation and traditional methods of water and wastewater treatment or between tasks performed by humans and those performed by machines (e.g., a policeman directing traffic versus the use of traffic control devices). Finding a balance has far-reaching implications for making a city caring or a city manageable. Thus, a totally automated city, which is technically possible, would also be an inhuman city. The balance between humans and other species will determine the extent to which a city favors biological diversity—the plants and animals that enrich the human environment.

Within the biosoma paradigm, trade-offs among information, materials, and energy are central to the concept of "intelligent" infrastructure, such as intelligent highways that can accommodate more traffic without the construction of new roadways. Trade-offs between materials and energy range from a simple, but ecologically significant, trade-off between using insulation instead of active heating and cooling to the utopian concept of a domed city, which uses material structures to control climate and, therefore, energy expenditures, but is unworkable for a variety of reasons. The trade-offs between biological and machine energy affect the extent to which walking or bicycling replaces motorized means of transportation, an important consideration in the design of cities as clusters of neighborhoods.

Biosomic cities based on balances and trade-offs will continually evolve. Society will continue to be transformed by the push and pull of innovations; and the machine component will continue to change with advances in information and telecommunications technology and the development of new materials, energy technologies, and methods of construction, reconstruction, and recycling. As each component of the biosoma changes, the balance among them must also change.

The emerging knowledge city and ecoindustrial city are embryonic manifestations of the biosomic city of the future (Figure 5). In the knowledge city, the emphasis for each component of the biosoma is on knowledge and information: in the biological component, on learning and biotechnology; in the social component, on education and e-business; and in the machine component, on computers, telecommunications, and nanotechnology.

One instrument of the knowledge city, the knowledge park, is congruent with the concept of neighborhoods and clusters. The knowledge park coalesces socioeconomic activities around institutions that generate knowledge (e.g., universities or research centers), transmit knowledge (e.g., schools), and use knowledge (e.g., business or industry and government). Knowledge parks will be increasingly important to the socioeconomic development of a knowledge society and will attract other elements of the city's organization and infrastructure (Figure 6).

The knowledge park provides a new organizing principle for the knowledge city by transforming the urban environment and providing an enormous economic boost. A case in point is Metrotech, catalyzed in Brooklyn, New York, by Polytechnic University, which has attracted some 20,000 jobs around the university, mostly in information technology and telecommunications, and has revitalized a significant part of downtown Brooklyn (Bugliarello, 1996). Although the need for face-to-face interactions will continue, an increasingly important aspect of the evolving knowledge city is virtuality—the ability to conduct business transactions and other social interactions at a distance.

In the ecoindustrial city, the waste created by one industry becomes the input for another. In addition, the biological and machine components are integrated and support each other (e.g., bioremediation of polluted areas). A pioneering example of this integration is the Danish city of Kalundborg (Graedel, 1999). Whatever shape the city of the future assumes, the challenge to planners, managers, and citizens will be to determine consciously the most desirable biological-social-machine balance.

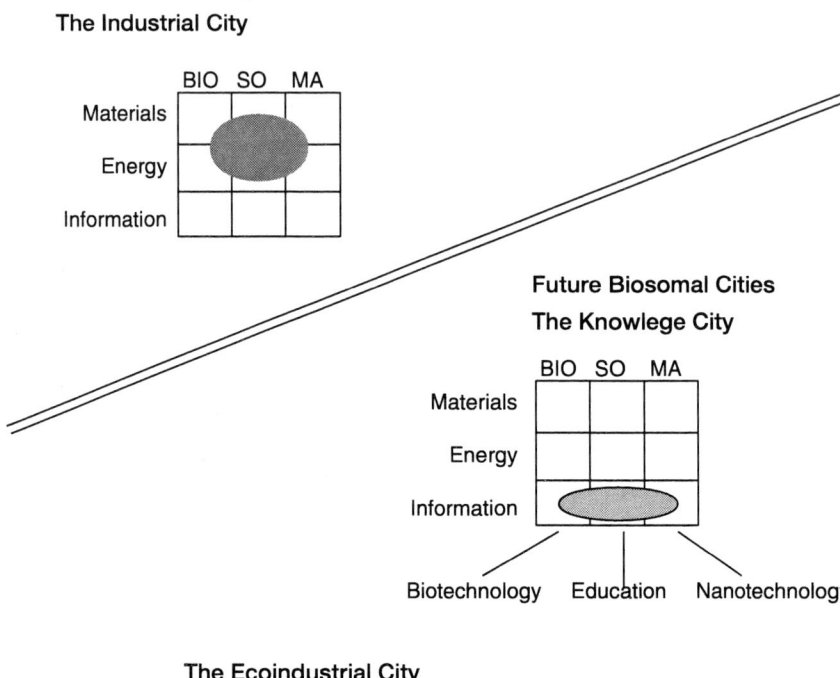

FIGURE 5 Comparison of past and future biosomic cities.

ENGINEERING CHALLENGES

The city of the future will present unprecedented engineering challenges, such as maintaining internal conditions within acceptable limits as the city is exposed to changes in temperature, winds, floods, and earthquakes, as well as to anthropogenic disasters, such as war and terrorism. The overall engineering challenge will be to limit the effects of these changes through design and operational decisions. For instance, although a city totally covered by a dome is unrealistic, a skyline—the location and configuration of structures—engineered to affect temperature and wind patterns is entirely feasible. A second challenge will be to minimize the effects of the city—its wastes and noxious emissions—on its surroundings, such as watersheds. A third challenge will be to develop technologies that address problems at the microscale of the neighborhood or the individual home, such as in-house energy transformers, waste disposal and recycling systems, and virtual offices. In appropriate situations, these technologies could provide alternatives to macroscale technologies, such as trunk utilities and other centralized services.

MOVING FORWARD

Transforming today's dysfunctional cities into tomorrow's less dysfunctional ones will, of course, require resources. But the will to transform the city will be even more important and generally more difficult to mobilize. The fundamental

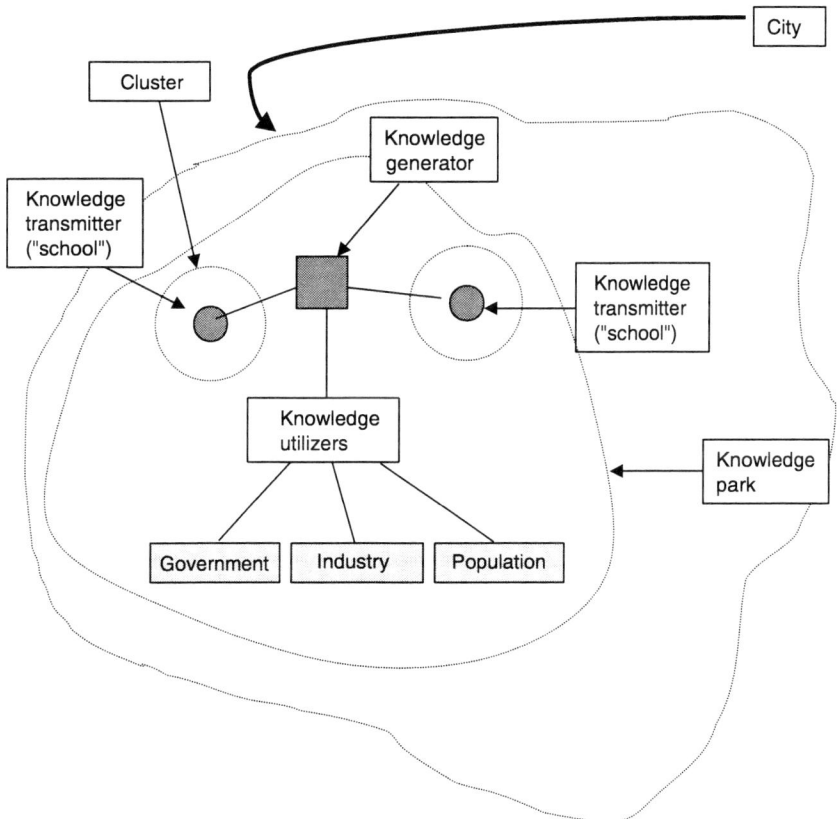

FIGURE 6 The knowledge city.

instrument for generating the will to change will be education that promotes reasonable expectations and explores ways they can be realized. Education must also encourage citizen participation in decision making and a willingness to make short-term sacrifices for the sake of long-term benefits. The city as a whole must be willing, when necessary, to accept temporary economic losses to ensure a more sustainable future.

Current trends strongly suggest that cities of the future will be home to an increasing share of the world population. We do not know, however, when the saturation point will be reached or whether urban populations will eventually decline. Nor do we know if cities of the future will be more dense and compact or more spread out than cities today. Despite these uncertainties, we already have much of the knowledge and technology we need to make future cities more effective, less dysfunctional instruments of human advancement, and we can expect new technologies to strengthen this capability (Ausubel and Herman, 1988). These technologies must be developed and applied in the context of a vision of the city that is caring and emotionally satisfying, ecologically sound, intelligent, and manageable. Given the rapid pace of urbanization and the level of dysfunction of many cities today, we must begin to address these problems immediately.

REFERENCES

Ausubel, J.H., and R. Herman, eds. 1988. Cities and Their Vital Systems: Infrastructure Past, Present, and Future. Washington, D.C.: National Academy Press.
Brennan-Galvin, E. 2000. What are the demographic trends? Presentation at the Megacities Workshop, National Research Council, Washington, D.C., September 26, 2000.
Bugliarello, G. 1996. Urban knowledge parks and economic and social development strategies. Journal of Urban Planning and Development 122(2): 33–45.
Bugliarello, G. 1998. Macroscope: Biology, society and machines. American Scientist 86(3): 230–232.
Bugliarello, G. 2000. The biosoma: the synthesis of biology, machines, and society. Bulletin of Science, Technology and Society 20(6): 452–464.
Craig, T. 1988. Air traffic congestion: problems and prospects. Pp. 222–231 in Cities and Their Vital Systems: Infrastructure Past, Present, and Future, J. H. Ausubel and R. Herman, eds. Washington, D.C.: National Academy Press.
Gallopin, G. C., S. Funtowicz, M. O'Connor, and J. Ravetz. 2001. Science for the 21st century: From social contract to the scientific core. International Journal of Social Science 54(168): 219–229.
Graedel, T.E. 1999. Industrial ecology and the ecocity. The Bridge 29(4): 10–14.
Groat, C. 2000. Crosscutting issue #1: Can we measure the urban footprint? What is the influence of the megacity beyond its official boundary? Presentation at the Megacities Workshop, National Research Council, Washington, D.C., September 26, 2000.
Huntley, E. 2001. Ground Zero. Photos available online at <http://www.geocities.com/cas43.geo/page2.html>.
Istituto e Museo di Storia della Scienza. 2001. Studies of Sforzinda, the ideal city projected by Leonardo for Ludovico il moro. Leonardo da Vinci Ms. B. Available online at <http://galileo.imss.firenze.it/news/mostra/6/ e62msb5.html> (April 18, 2001)
Lewis, P.H. 1998. Tomorrow by Design. New York: John Wiley.

Moss, M. 1998. Technology and cities. Cityscape: A Journal of Policy Development and Research 3(3): 107–127.
Perlman, J. 2000. Comment made during a discussion at Megacities Workshop, National Research Council, Washington, D.C., September 26, 2000.
Perry, C.A. 1929. The neighborhood unit: a scheme of arrangement for the family-life community. Pp. 2–140 in Neighborhood and Community Planning, Vol. VII of the Regional Study of New York and Its Environs. New York: Port of New York Authority.
Relph, E. 1987. The Modern Urban Landscape: 1880 to the Present. Baltimore: Johns Hopkins University Press.
Rowland, S.F. 2000. Can megacities be green? environmental issues in megacities. Presentation at the Megacities Workshop, National Research Council, Washington, D.C., September 26, 2000.
Rykwert, J. 2000. The Seduction of Place: The City in the Twenty-First Century. New York: Pantheon Books.

Urban Design
The Grand Challenge

LAWRENCE T. PAPAY

In February 2000, as part of Engineers' Week, I addressed the D.C. Council of Engineering Societies. In my talk, I referred to the National Academy of Engineering's (NAE's) list of the top 20 engineering achievements of the twentieth century and predicted what I thought the engineering challenges in the twenty-first century would be, especially (1) the development of complex systems, particularly as they apply to infrastructure, and (2) sustainability. These two challenges are components of a greater challenge—urban design.

If we consider the 20 greatest achievements of the past century (Table 1), we are struck by the number that deal with processes, services, and conveniences we associate with urban living. These include electrification, the automobile,

TABLE 1 The 20 Greatest Engineering Achievements of the 20th Century

1. Electrification	11. Highways
2. Automobile	12. Spacecraft
3. Airplane	13. Internet
4. Water supply and distribution	14. Imaging
5. Electronics	15. Household appliances
6. Radio and television	16. Health technologies
7. Agricultural technologies	17. Petrochemical mechanization
8. Computers	18. Laser and fiber optics
9. Telephone	19. Nuclear technologies
10. Air conditioning and refrigeration	20. High-performance materials

Source: NAE, 2000.

agricultural technologies, water supply and distribution, and health technologies. These achievements have certainly improved the quality of life for everyone in the twentieth century. In addition, they have accelerated the massive global shift of populations to urban centers. In this country, 50 percent of the population was engaged in farming at the turn of the previous century in contrast to 2 percent today. A century ago, cities with a million inhabitants were rarities; today megacities of 10 million are becoming all too common, mostly in countries that are ill-equipped to handle these large concentrations of people.

Urban centers are an important focus area for Earth systems engineering because they have impacts that extend far beyond their urban centers. For example, large cities create their own microclimates through local changes in albedo (light reflection), heat generation, and humidity. And their waste streams can pollute local and regional bodies of water, even such large areas as the Bay of Bengal. Cities rely not only on technologies and infrastructural concepts of the twentieth century, but also on those of the nineteenth century. To compound the problem, cities, regions, and nations spend a good portion of their gross national products enlarging, modifying, and repairing this infrastructure without examining whether the fundamental design of urban systems is appropriate or whether new approaches might be more effective. As cities grow, the size, complexity, cost, and, scale of existing technologies used to transport goods and services must be reexamined. Urban centers should be designed to meet the needs of tomorrow.

As an example, let's look at water supply and distribution and their complements, wastewater and sewage. The provision of a potable water supply and reliable distribution have been major accelerants to the expansion of urban centers. About 200 years ago, as cities in the United States began to grow, they had to look beyond local ponds, wells, and cisterns for water supplies. In 1801, Philadelphia was the first U.S. city to install a water system; Cincinnati soon followed in the 1820s, New York in 1841, and Boston in 1848. By 1860 the 16 largest cities in the United States had installed a total of 136 water systems (NRC, 1984).

The realization of abundant (if not unlimited) water supplies enabled cities to switch technologies from the use of privy vaults, cesspools, and private sewers for wastewater and sewage disposal to a technology that used large amounts of water as the medium for diluting and transporting wastes beyond the city boundary. In the intervening years, the main supplement to these systems has been the treatment of wastewater streams before discharge.

In 1850, or even 1950, this approach might have been acceptable. But we must ask ourselves if it is the best approach for the twenty-first century megacity? If we could go back to square one and systematically examine the use of water and the disposal of wastes, would we choose to use existing technologies? Can we afford to use freshwater, our most critical resource, as a medium for diluting and transporting waste and then turn around and treat this high-volume stream to make it acceptable for discharge into the environment? Should we create more dual water systems? For example, should potable water be supplied in containers and existing water systems be converted to gray water only? Should we rely on chemical or biochemical treatments of waste treatment to reduce the need for freshwater as a dilutant and transport medium? And should we consider using modular, smaller-scale treatment systems complete with local recycling or integrate treatment systems into systems that meet other needs, such as electricity and heat?

I have used water as an example, but I could just as well have used power or communications or transportation. Large-scale networked infrastructures will not necessarily be a wise choice for future, megascale urban development in any of these arenas. The grand challenge for engineers designing future urban centers, then, is to understand the patterns of the supply and use of energy, materials, products, information, and services that underlie urban systems. This information would enable us to select (or develop) technologies on a scale appropriate for various activities and elements of infrastructure. In short, we must begin to think outside the box.

In tackling this grand challenge, engineers must address a number of key questions:

- Will more stringent environmental and safety requirements limit the technical options in certain economic sectors but increase the options in other sectors?
- Should we look for alternative methods of waste disposal or recycling?
- What is the preferable scale for recycling various commodities and materials for different elements of urban infrastructure? For example, should water recycling systems be implemented on a building/neighborhood scale, and if so, should this be done vertically or horizontally? That is, should water cascade through systems at poorer and poorer quality (i.e., vertical use), or should it be recycled for uses that require equivalent water quality (i.e., horizontal use)?
- Can we capture and reuse waste heat and chemical energy?
- Is industrial agriculture environmentally efficient compared with, say, "apartment complex gardening systems"?
- Should we focus on technologies for power and communications that can provide reliable service without massive, hard-wired infrastructures?
- Does "homeland defense" require new paradigms for urban infrastructure?

Introducing new technologies that are likely to disrupt old infrastructures is much more difficult in older urban areas than in developing countries that have less existing infrastructure. Existing cities can be changed only incrementally, because inertia is difficult to overcome, particularly for capital-intensive systems. We know how to optimize engineered systems, but if we do not consider their human dimensions, we will only repeat the failures of the past (e.g., the construction of low-income housing projects in U.S. inner cities). We must look to small-scale examples that have worked, such as Brooklyn Heights and New England villages where self-sufficiency and self-containment have worked together. If we begin to address these problems now, our successors in 2100 will note that in the year 2000, we began a quest to use Earth systems engineering to make significant, beneficial changes in urban design.

REFERENCES

NAE (National Academy of Engineering). 2000. Greatest Engineering Achievements of the 20th Century. Available online at <http//www.greatestachievements.org>.

NRC (National Research Council). 1984. Perspectives on Urban Infrastructure. Washington, D.C.: National Academy Press.

Hybrid Cities
A Basis for Hope

GEETA PRADHAN AND RAJESH K. PRADHAN

Some years ago a problem shared by many cities manifested itself in such a dramatic way that even the skeptics among skeptics sat up with alarm. The people of Mexico City realized with a horror that their city was sinking. Water drawn over the years to sustain city life had far exceeded the amount that trickled down to replenish underground sources, triggering dramatic subsidence at times. Excessive paving had made matters worse, causing water run-off, flooding, aquifer depletion, and reliance on an expensive water supply system. Mexico City illustrates starkly how unsustainable our current urban practices are.

In this paper we offer an alternative approach to the development of cities inspired by the Italian writer-philosopher Italo Calvino. In *Invisible Cities* (1986), Calvino describes the empire of the Tartar emperor Kublai Khan, which is crumbling. Khan is devastated. To divert him, the Venetian traveler Marco Polo recounts stories about the cities he has seen during his travels. He describes cities of memories, cities of dreams, thin cities and wide cities, trading cities, cities of desire, signs, and eyes, cities of names, and hidden cities. Soon it becomes clear to Khan that each of these fantastic places is really the same place—Kublai Khan's empire. But the down-in-the-dumps Khan sees no hope of getting out of the ever closing-in inferno. Polo tells him:

> There are two ways to escape suffering it. The first is easy for many: accept the inferno and become such a part of it that you can no longer see it. The second is risky and demands constant vigilance and apprehension: seek and learn to recognize who and what, in the midst of the inferno, are not inferno, then make them endure, give them space.

We give space in the following pages to an idea anchored in the wisdom of the past and the vast literature on administrative decentralization. We take into account some scientific and technological advances, tempering the grandeur and visions of utopia with the realization that the world's finite resources cannot keep pace with human activity and population growth. Like the spaces Calvino allows within the inferno, we try to give legitimacy and room for the development of small trends and innovative concepts in response to the problems of urbanization. In the end, we hope to offer a new model of development, a hybrid approach that combines the best of rural and urban attributes to create "a village in a city, a city in a village." Metaphorically, this model will encourage us to look beyond cities and rethink our urban centers as we design the cities of tomorrow. It will suggest subduing the inferno and diffusing pressures within megacities by bringing, as it were, the countryside in.

The world population, which reached 6.1 billion in mid-2000, is expected to increase to 8.1 billion by 2030 (United Nations, 2001). Projections show that almost all of this growth will be concentrated in urban areas of the less developed world. Rural-to-urban migration and the transformation of rural settlements into cities are expected to be key contributors to this trend. Although an increasing share of the world population lives in urban areas, the percentage of people living in very large urban agglomerations—called megacities—is still small. In 2000, 4.3 percent lived in cities of 10 million or more; by 2015, the number is expected to rise to 5.2 percent. Cities of 5 to 10 million inhabitants, which currently account for 2.6 percent of world population, will hold about 3.5 percent by 2015. By comparison, the number of people living in smaller cities, although increasing at a slower pace, is considerably larger. In 2000, 28.5 percent of the world population was living in cities of one million or less; by 2015, cities of this size will account for 30.6 percent of total population.

Cities, which account for just 2 percent of the world's surface, use a disproportionate amount of the world's resources. For instance, they produce roughly 78 percent of carbon emissions from the burning of fossil fuels and the manufacture of cement; 76 percent of industrial wood is used in urban areas. Some 60 percent of the planet's water tapped for human use goes to cities in one form or another (O'Meara, 1999).

Cities account for a majority of the world's wealth and provide more than 50 percent of the world's employment. If population growth remains on its current trajectory, the global workforce will swell from about 3 billion today to nearly 4.5 billion by 2050 (World Resources Institute, 2000). In a desperate search for jobs, higher income, and more options, people will continue to be drawn to cities.

However, many urban environments are inhospitable and create incentives for people to move away and escape city life. Congestion, health risks related to pollution, ungovernability, and social chaos are common problems in some of the world's largest cities. According to the World Resources Institute (1996), at least 220 million people in cities of the developing world lack clean drinking water;

420 million do not have access to the simplest sanitation; and between one-third and one-half of city trash is not collected, contributing to flooding and the spread of disease. Domestic and industrial effluents released with little or no treatment into waterways affect the quality of water far beyond cities, rendering many urban rivers, for example, the Pasig River in Manila and the Yamuna River in New Delhi, biologically dead. Breakdowns and undercapacity in the aging infrastructure of cities, especially water supply and sewer systems, increase the incidence of waterborne and water-related diseases. At any given time, close to half the urban population suffers from one or more of these diseases (World Bank, 2000).

Rising rates of automobile ownership and the absence of public transportation and environmentally sound rapid transit systems have led to unprecedented levels of pollution and traffic congestion in cities. Urban air pollution is estimated to be responsible for more than three million deaths annually worldwide, almost all of them children (World Health Organization, 1997). The air in some cities in Latin America, China, and India has concentrations of pollutants, such as nitrogen oxide, sulphur dioxide and particulates, that are two to four times the safe levels set by the World Health Organization (Davis, 1999). The amount of air pollution children in these cities are exposed to is equivalent to their smoking two packs of cigarettes per day (World Bank, 2000).

Vehicle exhaust, the dominant ingredient in urban air pollution, is also spewing lead into the air. This toxic metal impairs the kidneys, liver, and reproductive system, and at high levels causes irreversible brain damage. A 1990 study of atmospheric lead pollution in Bangkok estimated that 30,000 to 70,000 children risked losing four or more points on their IQ levels because of high lead concentrations, and many more risked smaller decreases in intelligence (World Bank, 2000). Recent studies suggest that about two-thirds of children in New Delhi and an even greater proportion of children in Shanghai have blood levels of lead higher than the levels estimated to cause adverse health effects. In Cairo in early 1999, traffic in the city's industrial areas contributed to atmospheric lead concentrations that exceeded health guidelines by a factor of 11 (O'Meara, 1999).

Despite the problems associated with the growth of cities, development policies have continued to favor the urban sector. This "urban bias," to borrow a phrase popularized by the economist Michael Lipton in *Why People Stay Poor* (1977), is derived from a much earlier debate on how less industrialized nations should modernize. The strategy that gained acceptance, generally credited to Arthur Lewis (1954), was to focus on cities (as opposed to agricultural areas) as places that could provide jobs, produce goods at low wages due to surplus labor, generate wealth through exports and exchange, and create a dense environment that would encourage economic interdependence and innovation. The result of this development strategy, however, was excessive migration to cities, urban sprawl, and the relative stagnation of villages. Many cities, especially in the less industrialized world, have become unmanageable, ungovernable, and unsustainable.

In the United States and the rest of the industrialized world, the debate about cities is now framed differently. The focus is no longer on economic growth per se but on ways to improve the quality of urban life. These are new expressions of old ideas. Historically, American city planners saw the ideal city as one that took care of three aspects of human experience—work, family, and leisure. Over time, however, the ideal work life, family life, and leisure time have come to be associated with suburbs rather than inner cities. As family concerns about schools and safety in cities increase, employment and leisure activities have increased in the suburbs. The results of this out-migration, or suburban sprawl, have been a substantial loss of agricultural land and forests, urban disinvestment, an increase in transportation requirements, and an increase in suburban residential and commercial land use.

The change in America's population illustrates this trend. The urban population, including people living in suburbs, grew from 60 percent of the total population in 1950 to about 80 percent in 2000 (Ecological Cities Project, 2001). In the Boston metropolitan region, for example, more than 37 percent of open space was lost to urban sprawl in just 50 years (Pradhan and Kahn, 2000). The impact of increased automobile use has also led to a decrease in productivity. Drivers in 70 metropolitan areas spend an average 40 hours a year sitting in stalled traffic, resulting in wasted fuel and lost productivity that costs about $74 billion annually (O'Meara, 1999).

Concerns about health, productivity, and overall quality of life provide incentives for people to move away from cities. People with the means may choose to maintain two homes, one in the countryside where they can enjoy serenity, the other in the bustling city where they can enjoy diversity and cultural activities, work, and create wealth. This phenomenon of dual habitation, which is growing in the West and among affluent city dwellers in the developing world, is an indication of what many more people would do if they could.

As cities continue to expand, the pressures to manage water and energy resources, organize food production and distribution, and manage basic amenities, such as housing, transportation, and public health, will increase. Particularly in megacities, access to and control over these resources and services are increasingly becoming arenas of social conflict, raising demands for social justice and civic representation. Ironically, even if populations do not grow at the predicted rate, cities will still have to contend with the normal wear and tear on vital infrastructure and more discriminating, politicized inhabitants.

Considering the enormity of the problems facing cities, our responses have been timid. They have ranged from popularizing environmentally sound technical solutions (e.g., energy conservation equipment and pollution-prevention and cleanup technologies) to appealing to people to establish a deep (spiritual) bond with the environment and nature. However, projected growth in urban population and demand for services cannot be sustained by tinkering with technology or conventional city planning tools. And the awareness-raising strategy appears to disregard the basic human impulse to act on the basis of narrow self-interest. In other words, neither approach provides much hope for significantly improving the quality of life in today's sprawling cities and emerging urban centers.

What we need is an inspiring vision that provides a new direction for cities of the future. We need a utopia of sorts, a basis for hope, and a redefinition of what a city is. One such vision is the "hybrid city," the concentrated development of urban centers with populations of one million or less and the creation of countryside or small-town-like environments within large urban centers.

A FRAMEWORK FOR HYBRID CITIES

The hybrid city would be a sustainable community that emphasizes civic engagement, social justice, environmental soundness, and economic diversity. It is based on an understanding of factors that have lured people over the ages to cities and of the qualities of life people seek when they move to the countryside and small towns. We have attempted to provide a broad framework—rather than "quick fixes" and ad hoc solutions—for creating what we call "a village in a city, a city in a village." The hybrid city attempts to combine the best qualities of cities—diversity, density, innovation, economic mobility, and access to means for human development—with the best qualities of villages or small towns—cultural wisdom, frugality, conservation, resource efficiency, a sense of scale and place, self-reliance, and a sense of community and connectedness.

The vision is based on lessons learned from innovations in food production, the creation of open space, waste management, and transportation, which were adopted to take the heat off the "infernos" that many large cities have become. These trends and innovations also offer hope for the sustainability of smaller urban centers. A few examples should suffice.

Village-like Self-reliant Activities in Cities

Many small-scale efforts to enhance urban sustainability or livability have successfully provided residents with goods and services produced locally. They are guided by the principle of self-reliance, a characteristic typically associated with the village or country town of the past, when transportation options were limited. Chinese cities, for instance, have long reserved surrounding areas for agriculture and used city-generated wastes to fertilize the fields, which, in turn, have met a significant portion of city demand for vegetables, meat, and poultry. In Africa, urban agriculture is a survival strategy for the poor (O'Meara, 1999). In Boston, 150 community gardens augment the food budgets of families in the inner city; the gardeners are often low-income families and the elderly. Urban gardeners in New York City are organizing to protect their ad-hoc urban gardens. The popularity of public markets that stock locally grown produce and food products is testimony to the latent demand for urban agriculture.

Hybrid cities would make urban agriculture an explicit element of city planning. To the extent that this would create a variety of jobs in production, processing, and support industries (favoring less-skilled workers), the strategy would also further the goals of equity and social justice. From an architectural or urban design point of view, urban agriculture would enhance diversity in cities.

Village-like Open Spaces and Clean Air

Perhaps inadvertently, urban agriculture also provides badly needed open space in congested cities. Some U.S. metropolitan centers, such as Portland, Oregon, are working explicitly on ways to limit their boundaries, limit growth, and increase countryside-like open spaces. By moving a major highway underground, for instance, Boston is creating huge open spaces in the heart of its downtown. To reduce air pollution and create a more pristine environment, Chattanooga has replaced automobile traffic in the downtown area with free public transportation that runs on nonpolluting fuels (World Resources Institute, 2001). The change has led to massive economic investments in the city center.

Village-like Frugality and Resource Conservation

Curitiba, Brazil, has linked its waste recycling program to its efforts to boost nutrition. For every bag of recyclables citizens turn in, they receive a bag of locally grown vegetables. Similar recycling strategies are being used on an industrial scale. For instance, in Kalundborg, Denmark, waste from one industry feeds directly into another as raw material in a kind of "industrial symbiosis." Metropolitan Tokyo, with more than 80 percent of its land covered by asphalt, is harvesting rainwater for nondrinking uses by placing tanks on rooftops (O'Meara, 1999). Boston is conserving its drinking-water resources by replacing leaky pipes, installing water-saving features, and educating the public about the importance of water conservation. The city has reduced water loss in the past two decades from 33 percent to about 11 percent (Pradhan and Kahn, 2000).

City in a Village

With technological advancements, combined with traditional wisdom, we can create an island of city life surrounded by a sea of countryside. Anna Hazare's Raley Gaon Siddhi village project in Maharashtra, India, is one example (Hazare, 1997). The project has been hailed as one of the most successful sustainable community projects in India and has been replicated in more than 600 villages. The idea behind the Raley Gaon Siddhi project is not to create an urban center but to create a sustainable village with town-like diversity that provides a range of jobs and uses low-cost, environmentally sound technologies and watershed management to sustain village life.

Behind many of these innovative approaches to urban sustainability is the unspoken idea of providing cities with the qualities associated with life in the countryside. With technological advancements, city planners can create environments that respond simultaneously to the longing for the intensity of city life and the ideal of small-town life in a global economy. The "city in a village" can focus attention on the development of rural areas and small towns, "potential cities," as a way addressing the problem of unsustainable concentrations of population and economic activities in large cities.

HYBRID CITIES, DIVERSITY, AND VOICE

Another way of looking at hybrid cities is to focus on the issue of diversity, one of the defining characteristics of cities. The kinds of innovations we have just described will lead to a broadening and deepening of diversity. The coexistence of different ideas, opportunities, and experiences would create the conditions for constant innovation and creativity. The hybrid city would be both an actor in the global economy and a self-reliant entity that meets the needs of the local population for basic goods and services. Its diverse economy would be both industrial and craft-based, high tech and low tech, formal and informal.

Ideally, a hybrid city would be relatively small, governable, and manageable. It would provide a sense of community and allow people to feel connected to each other and to their city, thus building social capital and encouraging civic involvement. The hybridization of an existing megacity could occur in a number of ways, some of them complementary. First, a megacity could be broken up administratively into several small towns. Second, planning could be much more community-based or neighborhood-based, consistent with the decentralized units. Third, countryside-like spaces and activities could be incorporated along the periphery of a city, as well as in the city itself. Similarly, one can imagine high-technology-based urban clusters in the countryside. The idea, in other words, is to diversify both the city and the countryside.

To put it differently, the small town (or the countryside) should be a planning tool for the development of existing large cities. Because a hybrid city would incorporate ideas, such as direct political representation, inherent in small towns and rural settings, it would encourage civic engagement and social justice, which are critical to making cities more sustainable but are often lost in the rush to make cities more modern or more manageable. By advocating village-like or craft-like activities in production, processing, manufacturing, and services, the hybrid city would attempt to create a variety of employment opportunities and outlets for many skills, such as crafts and manufacturing, that are becoming irrelevant in the urban economy.

A CONGLOMERATION OF SMALL TOWNS

By conceptualizing the big city as a conglomeration of small towns interspersed by pockets of countryside, resource allocation and city planning would necessarily become neighborhood-based or community-based activities. It would facilitate civic engagement by relying on small administrative units as opposed to the centralized administration of traditional megacities. Thus hybrids would decentralize power and legitimize many different voices. Finally, by drawing attention to small urban centers and developing urban clusters within villages—possibly the hybrid cities of the future—investments would be directed to relatively forgotten communities.

The idea of enjoying the best of both worlds is not new. When conditions in cities became unbearable after the Industrial Revolution, urban thinkers developed utopian vision combining the best technologies with ideas of social justice to create equitable societies in harmony with nature. Whether or not we agree with them, ideas from Ebenezer Howard's garden cities movement, Le Corbusier's skyscrapers set in open parkland, and Frank Lloyd Wright's suburban sprawl made possible by the automobile, found their way into twentieth century urban planning throughout the world. Such is the power and influence of visions!

Unlike some of the utopian visions of the past, the hybrid city approach does not pretend to be a fully developed idea. It aims simply to unify disparate and badly needed attempts to ensure sustainability by mixing, like an alchemist, seemingly opposite elements—the city and the countryside, the megacity and the small town.

REFERENCES

Calvino, I. 1986. Invisible Cities. Translated by W. Weaver. Orlando: Harcourt Brace.

Davis, D. 1999. Urban Air Pollution Risks to Children: A Global Environmental Health Indicator. Washington, D.C.: World Resources Institute.

Ecological Cities Project. 2001. Changes in U.S. Metropolitan Populations and Land Area (1950–2050). Available online at <www.umass.edu/ecologicalcities/landusechange2.html> (March 27, 2001).

Hazare, A. 1997. A Veritable Transformation, Anna Hazare. Translated by B.S. Pendse. Raley Gaon Siddhi, Maharashtra, India: Raley Gaon Siddhi Pariwar Publications.

Lewis, W.A. 1954. Economic Development with Unlimited Supplies of Labour. Manchester School of Economic and Social Studies 22(2): 139–191.

Lipton, M. 1977. Why People Stay Poor: Urban Bias in World Development. Cambridge, Mass.: Harvard University Press.

O'Meara, M. 1999. Reinventing Cities for People and the Planet. Worldwatch Paper 147. Washington, D.C.: Worldwatch Institute.

Pradhan, G., and C. Kahn. 2000. The Wisdom of Our Choices: Boston's Indicators of Progress, Change, and Sustainability. Boston: The Boston Foundation.

United Nations. 2001. World Population Prospects: The 2000 Revision. New York: United Nations.

World Bank. 2000. Entering the 21st Century. World Development Report 1999/2000. Washington, D.C.: World Bank Publications.

World Health Organization. 1997. Conquering Suffering, Enriching Humanity. World Health Report. Washington, D.C.: World Health Organization.
World Resources Institute. 1996. World Resources 1996–1997 Guide to the Global Environment: The Urban Environment. Washington, D.C.: World Resources.
World Resources Institute. 2000. Global Trends. Washington, D.C.: World Resources Institute. Available online at <www.igc.org/wri/powerpoints/trends/sld001.html> (March 27, 2001).
World Resources Institute. 2001. Air Pollution: Chattanooga, Tennessee. Available online at <www.wri.org/enved/suscom-chattanooga.html> (April 18, 2001).

Appendixes

A

About the Authors

BRADEN R. ALLENBY is the environment, health, and safety vice president for AT&T and an adjunct professor at Columbia University, Princeton Theological Seminary, and the University of Virginia. From 1995 to 1997 he was director for energy and environmental systems at Lawrence Livermore National Laboratory, and in 1992 he was the J. Herbert Hollomon Fellow at the National Academy of Engineering. Dr. Allenby graduated from Yale University in 1972 and holds a J.D. from the University of Virginia Law School, an M.A. in economics from the University of Virginia, and an M.S. and Ph.D. in environmental sciences from Rutgers University. Dr. Allenby is a member of the Advisory Committee of the United Nations Environmental Programme's Working Group on Product Design for Sustainability, the board of directors of the Environmental Law Institute, and the scientific advisory board of the DOE/DOD/EPA Strategic Environmental Research and Development Program. He is on the editorial boards for the *Journal of Industrial Ecology*, the *Bulletin of Science, Technology and Society*, and the *Journal of Sustainable Product Design*. Dr. Allenby has written extensively on industrial ecology and design for the environment. He authored the first industrial ecology policy textbook and coauthored the first engineering textbook on industrial ecology. He is the Batten Fellow, Darden Graduate School of Business, University of Virginia; a member of the Virginia Bar; and a fellow of the Royal Society for the Arts, Manufactures and Commerce.

GEORGE BUGLIARELLO is the chancellor and former president (1973–1994) of Polytechnic University in Brooklyn, New York, and holds an Sc.D. degree from MIT. A past president of Sigma Xi, the scientific research society, Dr. Bugliarello is a founding fellow of the American Institute of Medical and Biological Engineering and a member of the National Academy of Engineering and the Council on Foreign Relations. Dr. Bugliarello has chaired the Board of Science and Technology for International Development and the Board on Infrastructure and Constructed Environment of the National Research Council. His international experience includes consultancies abroad for UNESCO and OECD. He is the U.S. member of NATO's Science for Peace Steering Group, and was previously the U.S. member of NATO's Science for Stability Steering Group. He is founder and coeditor of *Technology in Society: An International Journal* and the interim editor in chief of *The Bridge*, a quarterly published by the National Academy of Engineering. Dr. Bugliarello has spearheaded the creation of Metrotech, the nation's largest urban university-industry park.

ROBERT W. CORELL is a senior research fellow with the Environment and Natural Resources Program at Harvard University's John F. Kennedy School of Government and a senior fellow in the Atmospheric Policy Program of the American Meteorological Society. Prior to January 2000, he was assistant director for geosciences at the National Science Foundation. Dr. Corell served as chair of the National Science and Technology Council committee that oversees the U.S. global change research program and has served as chair and principal U.S. delegate to international bodies with interests in research on climate and global change. Dr. Corell's research is focused on the sciences of global change and interfaces between science and public policy. He is chair of the steering committee for the Arctic Climate Impact Assessment, an international assessment of the effects of climate variability and change and increases in ultraviolet radiation. Dr. Corell is an oceanographer and engineer. He earned his B.S., M.S., and Ph.D. degrees from the Case Institute of Technology and MIT.

JOHN H. (JACK) GIBBONS served from February 1993 to April 1998 as assistant to the president for science and technology and director of the Office of Science and Technology Policy. Since leaving the White House, Dr. Gibbons has continued to be actively involved in a variety of public and private service activities, including the International Energy Panel of the President's Committee of Advisors on Science and Technology, the steering committee of the National Climate Assessment, and the Committee of Advisors of the National Renewable Energy Laboratory. He received a B.S. in mathematics and chemistry from Randolph-Macon College, a Ph.D. in physics from Duke University, and has been awarded six honorary doctorates. He has received numerous awards and published extensively in the areas of energy and environmental policy, energy supply and demand, conservation, technology and policy, resource and environmental management, nuclear physics, and the origins of solar system elements. He is currently a senior fellow at the National Academy of Engineering, special advisor to the U.S. Department of State, and president-elect of Sigma Xi, the scientific research society.

EDWARD A. HILER, vice chancellor for agriculture and life science, dean of the College of Agriculture and Life Science, director of the Texas Agricultural Experiment Station, and director of the Texas Agricultural Extension Service, provides overall leadership for the agriculture program of the Texas A&M University System. He earned his B.S., M.S., and Ph.D. in agricultural engineering at Ohio State University and is a licensed professional engineer in Texas. Dr. Hiler has served as a consultant to Congress, the U.S. Department of the Interior, and several universities throughout the United States and Europe on water conservation, environmental quality, energy from biological processes, and the future of agricultural engineering. He is a member of the National Academy of Engineering, a past president of the American Society of Agricultural Engineers, and a fellow of the American Association for the Advancement of Science and the Institution of Agricultural Engineers in England. Dr. Hiler is the author or co-author of more than 100 publications.

ANITA K. JONES is the Lawrence R. Quarles Professor of Engineering and Applied Science at the University of Virginia. From 1993 to 1997 she served as director of defense research and engineering for the U.S. Department of Defense (DOD), where she was responsible for management of the science and technology program, oversight of DOD laboratories, and was the principal advisor to the secretary of defense for defense-related scientific and technical matters. Professor Jones is currently the vice chair of the National Science Board, which advises the president on science and engineering and oversees the National Science Foundation. She is also a member of the Defense Science Board, the Charles Stark Draper Laboratory Corporation, and the Board of Directors of Science Applications International Corporation. She has received the Computing Research Association's Service Award, the Air Force Meritorious Civilian Service Award, and the Department of Defense Award for Distinguished Public Service. Professor Jones holds an A.B. from Rice University in mathematics, an M.A. in literature from the University of Texas, Austin, and a Ph.D. in computer science from Carnegie Mellon University.

JERRY M. MELILLO is a research scientist and codirector of the Ecosystems Center of the Marine Biological Laboratory in Woods Hole, Massachusetts. Dr. Melillo's research on biogeochemistry has focused on the global carbon cycle, the ecological consequences of tropical deforestation, and the impacts of climate change on land ecosystems. He founded the Marine Biological Laboratory's semester in environmental science, a program that brings undergraduates from small liberal arts colleges and universities to Woods Hole to work on environmental science. He holds a B.A. from Wesleyan University and a Ph.D. from

Yale University, has served as associate director for environment in the president's Office of Science and Technology Policy, and is currently directing an assessment for the federal government on the impacts of climate change.

NORMAN P. NEUREITER is science and technology advisor to the secretary of state. He has extensive experience in government and industry and a public policy background that includes close ties to academia. Since 1996 Dr. Neureiter has served as U.S. cochair of the U.S.-Japan Joint High Level Advisory Committee and as a U.S. commissioner of the Maria Sklodowska-Curie Joint Fund II, which supports cooperative science and technology research between U.S. and Polish scientists and engineers. From 1973 to 1996 Dr. Neureiter held a variety of positions at Texas Instruments (TI), including director of East-West business development, manager of international business development, vice president for corporate staff, director of TI Japan, and vice president of TI Asia. Prior to his work at Texas Instruments, Dr. Neureiter worked as international affairs assistant in the White House Office of Science and Technology Policy, served as deputy science attaché in the U.S. Embassy in Bonn, and was the first U.S. science attaché in Eastern Europe, based at the U.S. Embassy in Poland. Dr. Neureiter received a B.S. in chemistry from the University of Rochester and a Ph.D. in organic chemistry from Northwestern University.

LAWRENCE T. PAPAY, a nationally recognized authority on engineering, science, and technology, is a member of the National Academy of Engineering and serves on committees, panels, and task forces for numerous organizations, including the President's Council of Advisors on Science and Technology, the National Science Foundation, the National Research Council, the U.S. Department of Energy, the American Nuclear Society, and the Electric Power Research Institute. He is a registered professional nuclear engineer in California. Dr. Papay recently joined Science Applications International Corporation (SAIC) as sector vice president, where he is responsible for integrating technology and energy, environment, and information systems for governmental and commercial clients worldwide. Prior to joining SAIC, Dr. Papay was senior vice president and general manager of Bechtel Technology and Consulting, where he was responsible for monitoring new technologies and developing new businesses. His prior experience includes more than 20 years at Southern California Edison. Dr. Papay received a B.S. in physics from Fordham University and an M.S. and Sc.D. in nuclear engineering from MIT.

GEETA PRADHAN is director of the New Economy Initiative at the Boston Foundations, a former consultant in the field of sustainable community development and planning, and former director of Sustainable Boston, a city of Boston initiative. She is coauthor of the report, *The Wisdom of Our Choices: Boston's Indicators of Progress, Change, and Sustainability* (The Boston Foundation,

2000), serves on the boards of several nonprofit organizations in the United States, and is a trustee of the Social Enterprise Education Trust in India. Dr. Pradhan received her graduate degree in urban design from Harvard University and was a recipient of the 1999 Women Waging Peace Award, an international fellowship program at Harvard. She also received the 1999 Urban Edge Community Service Award in Boston and an Environmental Leadership Award in 1998 from the Environmental Protection Agency New England Region.

RAJESH K. PRADHAN works as a consultant in the fields of urban planning, social entrepreneurship, and urban governance. He has also worked at the Harvard Institute for International Development on strengthening the capacity of public and private institutions and nongovernmental organizations in small enterprise programs in developing countries. Mr. Pradhan received dual graduate degrees in urban studies and city planning and architecture studies from the Massachusetts Institute of Technology (MIT) and did his doctoral work in political science (dissertation pending) also at MIT. He is the author of journal articles on small enterprises and the informal sector in developing countries and has published poems online on philosophical themes. He founded and heads the Social Enterprise Education Trust in India, an institution committed to promoting education, research, and activism in social entrepreneurship as a way of improving urban governance, urban environment, and livelihood and employment opportunities in urban areas.

DANIEL R. SAREWITZ is a senior research scholar and the managing director of Columbia University's Center for Science, Policy, and Outcomes. His work focuses on strengthening the connections between scientific research and social benefits. He is the coeditor of *Prediction: Science, Decision-Making, and the Future of Nature* (Island Press, 2000) and the author of *Frontiers of Illusion: Science, Technology, and the Politics of Progress* (Temple University Press, 1996). His written work also includes an article in the *Atlantic Monthly* (July 2000) on global climate change. Dr. Sarewitz was formerly director of the Geological Society of America's Institute for Environmental Education. He also served as science consultant to the U.S. House of Representatives Committee on Science, Space, and Technology and principal speech writer for Committee Chair George E. Brown, Jr. Before moving into the policy arena, he was a research associate and lecturer in the Department of Geological Sciences at Cornell University. He received his Ph.D. in geological sciences from Cornell University in 1986.

MAXINE F. SINGER is president of the Carnegie Institution and scientist emeritus at the National Cancer Institute's Laboratory of Biochemistry, where she was chief from 1980 to 1987. Dr. Singer received her A.B. from Swarthmore College and her Ph.D. in biochemistry from Yale University. Her research contributions have ranged over several areas of biochemistry and molecular biology, including human transposable elements, the structure and evolution of defective viruses, and enzymes that work on DNA and its complementary molecule RNA. Throughout her career Dr. Singer has taken a leading role in refining the nation's science policy. She was an organizer of the 1975 Asilomar conference on recombinant DNA research and one of five signers of the summary statement that drew up guidelines for that research. Dr. Singer is a member of the Human Genome Organization and serves on the board of directors for Johnson & Johnson. She is also a member of the National Academy of Sciences and the Pontifical Academy of Sciences and has received honorary degrees from Dartmouth, New York University, Swarthmore, Harvard, Yale, and numerous other institutions. In 1998 Dr. Singer was awarded the Distinguished Presidential Rank Award and in 1992 the National Medal of Science, the nation's highest scientific honor, "for her outstanding scientific accomplishments and her deep concern for the societal responsibility of the scientist."

KATHLEEN C. TAYLOR is director of the Materials and Processes Laboratory of the General Motors Research and Development and Planning Center, where she manages research and development in materials science and oversees the research of 105 engineers and scientists on lightweight bodies and power trains, engineered and advanced functional materials, and materials integration. Dr. Taylor is concurrently chief scientist for General Motors of Canada, Ltd. She was elected to the National Academy of Engineering in 1995, is a member of the American Chemical Society and the Materials Research Society, and is a fellow of the Society of Automotive Engineers and the American Association for the Advancement of Science.

ROBERT M. WHITE was president of the National Academy of Engineering from 1983 until his retirement in June 1995. Currently, Dr. White is a principal of the Washington Advisory Group, LLC, a science, technology, and education consulting firm. He is also a senior fellow at the University Corporation for Atmospheric Research. He served under five U.S. presidents from 1963 to 1977, first as chief of the U.S. Weather Bureau and finally as the first administrator of the National Oceanic and Atmospheric Administration. In these capacities, he is credited with bringing about a revolution in the U.S. weather warning system with satellite and computer technology, helping to initiate new approaches to the balanced management of the country's coastal zones, and strengthening American fisheries. As U.S. representative to the World Meteorological Organization from 1963 to 1978, he helped establish the World Weather Watch for the continuous monitoring of the Earth's atmosphere, the Global Weather Experiment to extend the time range of weather forecasts, and the World Climate Program to improve our understanding of climate change. Dr. White holds a B.A. in geology from Harvard University and an M.S. and Sc.D. in meteorology from Massachusetts

Institute of Technology. He holds honorary degrees from many universities and is a member of the Academies of Engineering of the United Kingdom, Japan, Australia, Russia, and Finland, as well as the French Legion of Honor. His many awards include the Vannevar Bush Award of the National Science Board, the Tyler Prize for environmental achievement, the Charles E. Lindbergh Award for technology and environment, the Rockefeller Public Service Award for Protection of Natural Resources, the Smithsonian Institution's Matthew Fontaine Maury Award for Contributions to Undersea Exploration, the International Conservation Award of the National Wildlife Federation, and the International Meteorological Organization Prize of the World Meteorological Organization.

B

Symposium Agenda

National Academy of Engineering Technical Symposium
Earth Systems Engineering
October 24, 2000
National Academies Building Auditorium

Continental Breakfast available in the Great Hall starting at 7:30 am

8:45 am **Welcome**
John H. Gibbons (NAE 1994)—NAE Senior Fellow

9:00 am **Keynote Address**
Norman P. Neureiter, Science and Technology Advisor to the Secretary of State

Session One
Engineering to Understand, Adapt, and Mitigate Climate Change
9:30 am Introduction of panel
9:35 am Main speaker
Robert M. White (NAE 1968)—President Emeritus, NAE; Principal, Washington Advisory Group
10:00 am Short responses from
Robert W. Corell—Senior Research Fellow, Belfer Center for Science and International Affairs, Harvard Univerisity
Jerry M. Melillo—Codirector, Ecosystem Center, Marine Biological Laboratory
10:20 am Discussion (Q&A)
10:40 am Break

115

Session Two
Using Biological Activity for Humanity's Benefit
11:10 am   Introduction of panel
11:15 am   Main speaker
              Maxine F. Singer *(NAS 1979, IOM 1979)—President, Carnegie Institution*
11:40 am   Short responses from
              Braden R. Allenby—*Environment, Health, and Safety Vice President, AT&T*
              Edward A. Hiler *(NAE 1987)—Vice Chancellor for Agriculture and Life Sciences, Texas A&M University System*
12:00 am   Discussion (Q&A)
12:20–1:30 pm Lunch

Session Three
Engineers and Policy Makers: Partners in Developing and Implementing Solutions
1:30 pm   Introduction of panel
1:35 pm   Main speaker
              Anita K. Jones *(NAE 1994)—Lawrence R. Quarles Professor of Engineering and Applied Science, Department of Computer Science, University of Virginia*
2:00 pm   Short responses from
              Kathleen C. Taylor *(NAE 1995)—Director, Material and Processes Laboratory, Research and Development Center, General Motors Corporation*
              Daniel R. Sarewitz—*Managing Director and Senior Research Scholar, Center for Science, Policy, and Outcomes, Columbia University*
2:20 pm   Discussion (Q&A)
2:40 pm   Break

Session Four
Rethinking Today's Cities: Designing Tomorrow's Urban Centers
3:10 pm   Introduction of panel
3:15 pm   Main speaker
              George Bugliarello *(NAE 1987)—Chancellor, Polytechnic University*
3:40 pm   Short responses from
              Lawrence T. Papay *(NAE 1987)—Vice President for the Integrated Solutions Sector, SAIC*
              Geeta Pradhan—*Director, Sustainable Boston*

4:00 pm   Discussion (Q&A)
4:20 pm   Closing Remarks

4:30 pm   Adjourn—Reception in the Great Hall